D1116064

THE FRONTIERS COLLECTION

THE FRONTIERS COLLECTION

The books in this collection are devoted to challenging and open problems at the forefront of modern science, including related philosophical debates. In contrast to typical research monographs, however, they strive to present their topics in a manner accessible also to scientifically literate non-specialists wishing to gain insight into the deeper implications and fascinating questions involved. Taken as a whole, the series reflects the need for a fundamental and interdisciplinary approach to modern science. Furthermore, it is intended to encourage active scientists in all areas to ponder over important and perhaps controversial issues beyond their own speciality. Extending from quantum physics and relativity to entropy, consciousness and complex systems—the Frontiers Collection will inspire readers to push back the frontiers of their own knowledge.

More information about this series at http://www.springer.com/series/5342

For a full list of published titles, please see back of book or springer.com/series/5342

Victor Callaghan · James Miller
Roman Yampolskiy · Stuart Armstrong
Editors

THE TECHNOLOGICAL SINGULARITY

Managing the Journey

Editors
Victor Callaghan
School of Computer and Electrical Engineering
University of Essex
Essex
UK

Roman Yampolskiy
Department of Computer Engineering and Computer Science
University of Louisville
Louisville, KY
USA

James Miller
Economics Faculty
Smith College
Northampton, MA
USA

Stuart Armstrong
Faculty of Philosophy
University of Oxford
Oxford
UK

ISSN 1612-3018 ISSN 2197-6619 (electronic)
THE FRONTIERS COLLECTION
ISBN 978-3-662-54031-2 ISBN 978-3-662-54033-6 (eBook)
DOI 10.1007/978-3-662-54033-6

Library of Congress Control Number: 2016959969

Printed on acid-free paper

This Springer imprint is published by Springer Nature
The registered company is Springer-Verlag GmbH Germany
The registered company address is: Heidelberger Platz 3, 14197 Berlin, Germany

Foreword

The technological singularity is based on the ongoing improvements in and power of artificial intelligence (AI) which, at some point, will enable intelligent machines to design successive generations of increasingly more powerful machines, eventually creating a form of intelligence that surpasses that of humans (Kurzweil 2006). The technological singularity (or 'singularity' as I shall refer to it from hereon) is the actual point past which events are beyond the control of humans, resulting either in humans upgrading (with implants) to become cyborgs or with intelligent machines taking complete control. As well as being not particularly good news for ordinary humans, it is clearly an extremely critical point in time.

Already AI exhibits several advantages over the human form of intelligence. These are perhaps most easily witnessed in such terms as mathematical processing, memory, multidimensional operation, multi-sensory capabilities, body extension and, perhaps above all else, heightened forms of communication. As these advances accelerate, producing artificial general intelligence (AGI) whereby its future capabilities may well be impossible for humans to comprehend. This book seeks to give us some understanding of what lies ahead. In doing so, I believe the coverage is balanced, realistic and not at all sensationalist. Whilst the possibility of a major machine takeover is sensibly recognised by all authors, many of the articles look at ways of combating the situation or at least of how humans might live with intelligent machines rather than being ruled by them.

This particular work is therefore of considerable importance, containing, as it does, a focused collection of articles by some of the world's leading thinkers and experts on the topic. In this foreword, I do not wish to merely give a two-line precis of each entry but rather I have selected some of the works that caught my eye in particular. What follows is therefore my completely biased assessment.

Importantly, the two chapters by Yampolskiy and Sotala, which form Part I, take a serious look at what exactly AGI is all about and what the real risks are. They go on to consider in depth a wide variety of pro and con arguments with regard to the reality of these risks. These include giving a voice to those who feel that any dangers are (as they call them) over-hyped and we need to do nothing but we can all sleep soundly. In this part, some of the different possible outcomes of the singularity are described in detail, ranging from an intelligent machine takeover through human-into-cyborgs upgrades but also encapsulating the possibility of a direct merger involving biological brains in a technological body.

Part II examines various aspects of a potential singularity in greater depth. Soares' article, for example, looks at an approach to align the interests and direction of super-intelligent entities with those of humans, or rather how humans can realign their own thinking and mode of operation to potentially encompass the rise of intelligent machines. On an even more conservative note, Barrett and Baum consider ways in which, as they read it, the low probability of an artificial super-intelligence can be managed. In this way, the risk is acknowledged but, they argue, safety features can be built in and human progress can be steered appropriately.

Zheng and Akhmad consider, from a perspective of change agency theory, how the course of any singularity will be driven by socio-economic factors and natural human development. In particular, they map out the path of change. For us to realise that intelligent machines can not only learn and adapt their ways and goals, but also pass this on to the next generation is an important factor. This is picked up on nicely by Majot and Yampolskiy in their article on recursive, self-improving machines. They believe that due to natural limits, things may well not turn out as bad as some suggest. Indeed, their message is that the suggested ever-widening gap between machine and human intelligence will not be as problematic as was first thought and that any dangerous situation could be halted.

Part III is entitled reflections and contains, amongst other things, a reprint of Vinge's original article on the singularity. This is a sterling contribution that is in itself well worth reading. In many ways, it puts the rest of this book in perspective and makes one realise how relatively recent is the whole field of study on which this book is based. Looking back merely a decade, terms such as 'human enhancement' were very much aimed at individual enhancements, often bringing that individual back towards a human norm for someone who had lost an ability due to an accident or illness. Now, because of the available technology and the scientific experiments which have since taken place, we can readily explore and discuss the possibilities of enhancement beyond a human norm.

In terms of artificial intelligence, just as Alan Turing predicted over 60 years ago, '*the use of words and general educated opinion will have altered so much that one will be able to speak of machines thinking without expecting to be contradicted*' (Turing 1950). Turing also saw machine intelligence to be somewhat different to human intelligence, which we can now generally see it to be for all sorts of reasons. One consequence of this difference is that it can potentially outperform the human version, rather like an aeroplane, in many ways, outperforms a bird in terms of flying. The actual instantiations of artificial intelligence and a more powerful artificial general intelligence therefore form an area of extremely exciting study in which new discoveries are being made regularly. However, along with many positive uses comes the threat of the singularity with all its consequences for present day humanity.

It would probably be sensible for readers to take on board the messages contained in this book sooner rather than later. In different ways, humans are endowing certain machines with increasing intellectual abilities and deferring to them in an ever-increasing fashion. Unfortunately, humans will most likely not be aware when the singularity has been reached until it is too late. Some pundits have suggested that indicators such as the passing of the Turing test may well give us a clear sign. However, this has been shown to be false (Warwick and Shah 2016). Most likely, the singularity will happen before we realise it and before we can do anything about it. Rather than getting closer to the edge of a cliff, as we take each step forward we will feel that all is OK until we take one step too far and there will be no turning back—the singularity will be upon us.

Prof. Kevin Warwick
Deputy Vice Chancellor (Research)
Coventry University
Coventry, UK

References

Kurzweil, R., (2006) "The Singularity is Near", Duckworth.

Turing, A. (1950) "Computing machinery and intelligence". Mind, LIX, pp. 433–460. doi:10.1093/mind/LIX.236.433.

Warwick, K. and Shah, H (2016) "Passing the Turing Test does not mean the end of Humanity", Cognitive Computation, DOI: 10.1007/s12559-015-9372-6, February, 2016.

Acknowledgements

We are indebted to many people for the successful completion of this book. We should start by acknowledging the pivotal role played by **Amnon H. Eden, James H Moor, Johnny H. Soraker and Eric Steinhart**, the editorial team of the earlier volume of the Springer Frontiers Collection "*Singularity Hypotheses: A Scientific and Philosophical Assessment*", who's excellent work opened up the opportunity for this follow-on publication. In particular, we are especially grateful to **Dr Amnon Eden** who spotted the opportunity for this book and who wrote the original proposal to Springer before passing the project, most graciously, over to the current editorial team. We are also pleased to express our deep gratitude to the excellent support provided by the Springer team, especially **Dr Angela Lahee** Executive Editor, Physics, Springer Heidelberg, Germany for the highly proficient support and genuine enthusiasm she has shown to us throughout the production of this book. Without Angela's support this book would not have been possible. Likewise, we are grateful to Springer's Indian team, most notably, **Shobana Ramamurthy**, for overseeing the final stages of production. We should also acknowledge the **Creative Science Foundation** (www.creative-science.org) and their head of media communications, **Jennifer O'Connor**, for hosting the 'call for chapters' (and supporting information) on their website and promoting it throughout their network. Finally, we wish to express our deeply felt thanks to the authors who have generously allowed us to reproduce their articles in support of this book. In particular we wish to thank the authors of the blogs reproduced in Chapter 14 (*Singularity Blog Insights*) namely, **Eliezer Yudkowsky** (*Three Major Singularity Schools*), **Stuart Armstrong** (*AI Timeline Predictions: Are We Getting Better*), **Scott Siskind** (*No Time Like the Present for AI Safety Work*) and **Scott Aaronson** (*The Singularity is Far*). Last, but not least, we wish to thank **Vernor Vinge** for his permission to reproduce his visionary 1993 article "*The Coming Technological Singularity: How to Survive in the Post-human Era*" in the appendix of this book. By way of a concluding reflection, we should like to add that we are grateful to you, the reader, for taking the time to read this book and, as a consequence, to be part of a collective consciousness that we hope will influence any upcoming singularity towards an outcome that is beneficial to our world.

Victor Callaghan
James Miller
Roman Yampolskiy
Stuart Armstrong

Contents

1 **Introduction to the Technological Singularity** 1
Stuart Armstrong

Part I Risks of, and Responses to, the Journey to the Singularity

2 **Risks of the Journey to the Singularity** 11
Kaj Sotala and Roman Yampolskiy

3 **Responses to the Journey to the Singularity** 25
Kaj Sotala and Roman Yampolskiy

Part II Managing the Singularity Journey

4 **How Change Agencies Can Affect Our Path Towards a Singularity** 87
Ping Zheng and Mohammed-Asif Akhmad

5 **Agent Foundations for Aligning Machine Intelligence with Human Interests: A Technical Research Agenda** 103
Nate Soares and Benya Fallenstein

6 **Risk Analysis and Risk Management for the Artificial Superintelligence Research and Development Process** 127
Anthony M. Barrett and Seth D. Baum

7 **Diminishing Returns and Recursive Self Improving Artificial Intelligence** 141
Andrew Majot and Roman Yampolskiy

8 **Energy, Complexity, and the Singularity** 153
Kent A. Peacock

9 **Computer Simulations as a Technological Singularity in the Empirical Sciences** 167
Juan M. Durán

10 Can the Singularity Be Patented? (And Other IP Conundrums for Converging Technologies) 181
David Koepsell

11 The Emotional Nature of Post-Cognitive Singularities 193
Jordi Vallverdú

12 A Psychoanalytic Approach to the Singularity: Why We Cannot Do Without Auxiliary Constructions 209
Graham Clarke

Part III Reflections on the Journey

13 Reflections on the Singularity Journey 223
James D. Miller

14 Singularity Blog Insights 229
James D. Miller

Appendix 245

Titles in this Series 257

Chapter 1
Introduction to the Technological Singularity

Stuart Armstrong

The term "technological singularity" has been much used and abused over the years. It's clear that it's connected with the creation of artificial intelligences (AIs), machines with cognitive skills that rival or surpass those of humanity. But there is little consensus or consistency beyond that. In a narrow sense it can refer to AIs being capable of recursive self-improvement: of redesigning themselves to be more capable, and using that improved capability to further redesign themselves, and so on. This might lead to an intelligence explosion, a "singularity" of capability, with the AIs rapidly surpassing human understanding and slipping beyond our control. But the term has been used in a variety of other ways, from simply denoting a breakdown in our ability to predict anything after AIs are created, through to messianic and quasi-religious visions of a world empowered by super-machines (Sandberg's 2010, has identified nine common definition, corresponding to eighteen basic models).

It is in the simplest sense, though, that the term is the most useful. A singularity in a model—a breakdown of our ability to predict beyond that point. It need not mean that the world goes crazy, or even that the model does. But it does mean that our standard tools become inadequate for understanding and shaping what comes after. New tools are needed.

In a sense, the prevalence of robots and AI on our TV and computer screens have made us *less* prepared for the arrival of true AIs. Human fiction is written by humans—for the moment, at least—for humans, about human concerns, and with human morality and messages woven into them. Consider the potential for AIs or robots to take over human jobs. In fiction, the focus is variously on hard-working humans dispossessed by the AIs, on humans coping well or badly with the extra leisure (with a moral message thrown in one way or the other), or, maybe, on the robots themselves, portrayed as human-equivalents, suffering as slaves. On top of

S. Armstrong (✉)
Future of Humanity Institute, Oxford University, Oxford, UK
e-mail: stuart.armstrong@philosophy.ox.ac.uk

V. Callaghan et al. (eds.), *The Technological Singularity*,
The Frontiers Collection, DOI 10.1007/978-3-662-54033-6_1

that, all the humans (or human-equivalent robots) typically have the exact same cultural values as the writers of the work.

Contrast that with a more mechanical or analytic interpretation of the problem: what will happen when human capital (what we used to call "employees") can be mass produced as needed for any purpose by companies? Or by political parties, by charities, by think tanks, by surveillance organisations, by supermarkets, by parking services, by publishers, by travel agencies? When everything in the world that is currently limited by human focus and attention—which is most things, today—suddenly becomes a thousand times easier. Every moment of every day would be transformed, even something as simple as a lock: why bother with a key if a personal assistant can wait alongside or inside the door, opening as needed (and only as needed). On top of that, add the shock of (some? most? all?) jobs getting destroyed at the same time as there is a huge surge in manufacturing, services, and production of all kind. Add further the social dislocations and political responses, and the social and cultural transformations that will inevitably follow. It is clear that industry, economy, entertainment, society, culture, politics, and people's very conception of themselves and their roles, could be completely transformed by the arrival of AIs.

Though the point should not be belaboured (and there are certainly uncertainties and doubts as to how exactly things might play out) it is clear that our fiction and our conception of AIs as "yet another technology" could leave us woefully unprepared for how transformative they could be. Thus, "singularity" is not an inappropriate term: a reminder of the possible magnitude of the change.

This book is the second in a series looking at such an AI-empowered singularity. The first, "Singularity Hypotheses: A Scientific and Philosophical Assessment" looked squarely at the concept, analysing its features and likelihood from a variety of technical and philosophical perspectives. This book, in contrast, is focused squarely on managing the AI transition, how to best deal with the wondrous possibilities and extreme risks it may entail.

If we are to have a singularity, we want to have the best singularity we can.

1.1 Why the "Singularity" Is Important

Though the concept of machine intelligence is fascinating both from a storytelling and a philosophical point of view, it is the practical implications that are of highest importance in preparing for it. Some of these implications have already been hinted at: the possibility of replacing "human capital" (workers and employees) with AI-powered versions of this. While humans require a long, 18+ years of maturation to reach capability, and can perform one task at a time, an AI, once trained, could be copied at will to perform almost any task. The fallibilities of the human condition—our limited memories, our loss of skills due to ageing, our limited short term memories, our slow thinking speeds on many difficult and important problems, and so on—need not concern AIs. Indeed, AIs would follow a radically different learning cycle than humans: different humans must learn the same skill individually, while skills learnt by an AI could then become available for every AI to perform at comparable ability—just as once the basic calculator was invented, all computers could perform superhuman feats of multiplication from then on.

This issue, though not exactly prominent, has already been addressed in many areas, with suggestions ranging from a merging of humans and AIs (just as smartphone or cars are an extension of many people's brains today), through to arguments about AIs complementing rather than substituting for human labour, and radical political ideas to use the AI-generated surplus for guaranteeing a basic income to everyone. In a sense, the arguments are nothing new, variants of the perennial debates about technological unemployment. What is often underestimated, though, is the speed with which the change could happen. Precisely because AIs could have general intelligence—therefore the ability to adapt to new situations—they could be rapidly used in jobs of all varieties and descriptions, with whole categories of human professions vanishing every time the AI discovered a new skill.

But those issues, though interesting, pale in comparison with the possibility of AIs becoming superintelligent. That is the real transformative opportunity/risk. But what do we mean exactly by this superintelligence?

1.2 Superintelligence, Superpowers

One could define intelligence as the ability to achieve goals across a wide variety of different domains. This definition naturally includes the creativity, flexibility, and learning needed to achieve these goals, and if there are other intelligent agents in those domains, it naturally also includes understanding, negotiating with, and manipulating other agents.

What would a "super" version of that mean? Well, that the agent has a "super-ability" to achieve its goals. This is close to being a tautology—the winning agent is the agent that wins—so exploring the concept generally involves mining analogies for insight. One way to do so would be to look down: chimps and dolphins are quite intelligent, yet they have not constructed machinery, cities, or rockets, and are indeed currently dependent on human goodwill for their continued survival. Therefore it seems possible that intelligences as far "above" us as we are above chimps or cows, would be able to dominate us more completely than we dominate them.

We could imagine such an intelligence simply by endowing it with excellent predictive powers. With good predictions, draining all money from the stockmarkets becomes trivial. An agent could always know exactly how much risk to take, exactly who would win any debate, war, or election, exactly what words would be the most likely to convince humans of its goals, and so on. Perfect prediction would make such an AI irresistible; any objections along the lines of "but wouldn't people do X to stop it" runs up against the objection of "the AI could predict if people would try X, and would either stop X or find an approach so people wouldn't want to do X".

That superpredictor is almost certainly fanciful. Chaos is a very real factor in the world, reducing the ability of any entity to accurately predict everything. But an AI need not be able to predict *everything*, just to be able to predict enough. The stock market, for instance, might be chaotic to our human eyes, and might be mainly chaotic in reality, but there might be enough patterns for an AI to exploit. Therefore the concept of superintelligence is really an entwining of two ideas: that the AI could be much better than us at certain skills, and that those skills could lead to great abilities and power the world. So the question is whether there is a diminishing return to intelligence (meaning that humans will probably soon plumb the limits of what's possible), or is its potential impact extreme? And if the return to intelligence is limited, is that true both in the short and the long terms?

We cannot be sure at this point, because these questions are precisely about areas where humans have difficulty reasoning—if we knew what we would do with superintelligence, we'd do that already. Some have argued that, in a certain sense, humans are as intelligent as possible (just as all Turing-complete agents are equivalent), and that the only difference between agents is time, resources, and implementation details. Therefore, there is no further leap in intelligence possible: everything is on a smooth continuum.

Though that argument sounds superficially plausible, it can be undermined by turning it on its head and looking back into the past. It might be that there is an upper limit on social intelligence—maybe the most convincing demagogue or con man wouldn't be much better than what humans can achieve today. But scientific and technological intelligence seem different. It seems perfectly plausible for a superintelligence in 1880 to develop airplanes, car, effective mass production, possibly nuclear weapons and mass telecommunications. With skillful deployment, these could have made it superpowered at that time. Do we think that our era is somehow more resistant? Even if the curve of intelligence is smooth, it doesn't mean the curve of impact is.

Another analogy for superintelligence, created by this author, is of a super-committee. Consisting of AIs trained to top human level in a variety of skills (e.g. Einstein-level for scientific theorising, Bill Clinton-level for social interactions), and networked together perfectly, run at superhuman speeds, with access to all research published today in many domains. It seems likely to accumulate vast amounts of power. Even if it is limited over the short term (though recall its potential to be copied into most jobs currently done by humans), it's long-term planning could be excellent, slowly and inexorably driving the world in the direction it desires.

That is a potentially superintelligent entity constructed from human-level intelligences. Most of the other paths to superintelligence involves the AIs turning their intelligence towards developing their own software and hardware, and hence achieving a recursive feedback loop, boosting their abilities to ever greater levels (is such a feedback even loop possible? Add that question to the pile of unknowns). There is, however, another possibility for superintelligence: it is possible that the range of human intelligences, that seems so broad to us, is actually fairly narrow. In this view, humans are products of evolution, with some thin rational ability layered on top of a mind developed mainly for moving about in the savannah and dealing with tribal politics. And a small part of that rational ability was later co-opted by culture for such tasks as mathematics and scientific reasoning, something humans would be very bad at in any objective sense. Potentially, the AIs could develop at a steady pace, but move from being dumber than almost all humans, to radically smarter than everyone, in the space of a single small upgrade.

Finally, though we've been describing an AI becoming "superpowered" via being superintelligent, this need not be the case. There are ways that a relatively "dumb" AI could become superpowered. Various scenarios can be imagined, for an AI to dominate human society as it currently stands. For instance, a small AI with the ability to hack could take over a computer, copy itself there, take over two computers from its current basis, copy themselves there, and so on, thus taking over most of the internet in short order (a smart botnet). Alternately, there are ways that certain narrow abilities could lend themselves to AI takeover, such as weapon design and manufacturing (especially biological weapons such as targeted pandemics). A more thorough look at potential AI superpower can be found in Nick Bostrom's book (2014).

It seems therefore that there is a potential for extremely superintelligent AIs being developed, following soon after AIs of "human-level" intelligence. Whether or not this is possible depends on certain facts about the nature of intelligence, the potential return on new knowledge, and the vulnerabilities of human society, that we just can't know at the present.

1.3 Danger, Danger!

Of course, just because an AI could become extremely powerful, does not mean that it need be dangerous. Ability and morality are not correlated in humans (and even less so in the alien mind of an AI), so the AI could be extremely powerful while being extremely well-intentioned.

There are arguments to doubt that happy picture, however. Those arguments are presented mainly in the preceding book (Eden et al. 2012), and in other works such as Bostrom (2014) and Armstrong (2014), so the point won't be belaboured here too much. The risk is not that the AI could, by luck or design, end up "evil" in the human sense. The risk is rather that the AI's goals, while harmless initially, become dangerous when the AI becomes powerful. In fact, most goals are dangerous at high power. Consider the trivial example of a spam filter that becomes superintelligent. Its task is to cut down on the number of spam messages that people receive. With great power, one solution to this requirement is to arrange to have all spammers killed. Or to shut down the internet. Or to have everyone killed. Or imagine an AI dedicated to increasing human happiness, as measured by the results of surveys, or by some biochemical marker in their brain. The most efficient way of doing this is to publicly execute anyone who marks themselves as unhappy on their survey, or to forcibly inject everyone with that biochemical marker. We can't be sloppy when it comes to AI goal design—if we want "happiness", we have to fully unambiguously define it.

This is a general feature of AI motivations: goals that seem safe for a weak or controlled AI, can lead to extremely pathological behaviour if the AI becomes powerful. As the AI gains in power, it becomes more and more important that its goals be fully compatible with human flourishing, or the AI could enact a pathological solution rather than one that we intended. Humans don't expect this kind of behaviour, because our goals include a lot of implicit information, and we take "filter out the spam" to include "and don't kill everyone in the world", without having to articulate it. But the AI might be an extremely alien mind: we cannot anthropomorphise it, or expect it to interpret things the way we would. We have to articulate all the implicit limitations. Which may mean coming up with a solution to, say, human value and flourishing—a task philosophers have been failing at for millennia—and cast it unambiguously and without error into computer code.

Note that the AI may have a perfect understanding that when we programmed in "filter out the spam", we implicitly meant "don't kill everyone in the world". But the AI has no motivation to go along with the spirit of the law: its goals are the letter only, the bit we actually programmed into it. Another worrying feature is that the AI would be motivated to hide its pathological tendencies as long as it is weak, and assure us that all was well, through anything it says or does. This is because it will never be able to achieve its goals if it is turned off, so it must lie and play nice to get anywhere. Only when we can no longer control it, would it be willing to act openly on its true goals—we so we better have made those safe.

1.4 Uncertainties and Safety

The analogies and arguments of the preceding sections are somewhat frustrating. "It seems", "possibly", "it might", "maybe", and so on. Why is there so much uncertainty and so many caveats? Is everyone so vague about what AIs could be or do?

Certainly not. Vagueness is not the norm for AI predictions. Instead, the norm is for predictors to be bold, dramatic, confident—and wrong. We can comfortably say that most AI predictions are wrong, simply because they contradict each other. For a simple taste, see the following graph showing a variety of predictions for the dates at which we might have AI (for more details on this data see "AI timeline predictions: are we getting better?", see Chap. 14 in this book):

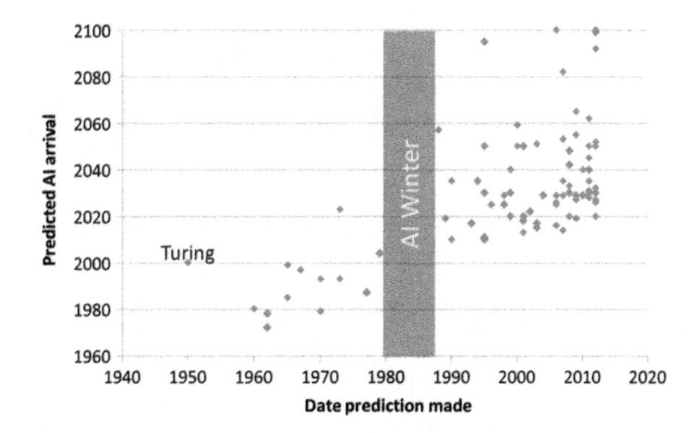

As can be seen, the predictions are all over the place, not showing any particular consensus, with wide intervals between them, and a few of the dates already passed. Digging into the details of the predictions—and the weak arguments that underpins the strong confidence—paints an even more dismal picture.

But, stepping back a bit, why *should* we expect anyone to be accurate about predicting AI? We don't have an AI. We don't know how to build one. We don't know what features it would have if it were built. We don't know whether consciousness, intuitions, and emotions would be necessary for an AI (also, we don't know what consciousness is). We don't know if it would need to be physically embodied, how fast it could run, whether it would make new discoveries at the speed of thought or at the slower speed of experiment, and so on. If predictions in politics are so poor—and they are, as Tetlock (2005) has demonstrated—and politics involve very understandable human-centric processes, why would we expect predictions in AI to be any better? We are indeed trying to predict a future technology, dependent on future unknown algorithms and the solutions to whole host of problems we can't yet even see.

There actually has been quite a bit of work on the quality of expert predictions. A lot of different analysis have been done, by such people as Shanteau et al. (2002), Khaneman (2011), Cooke (2004), Klein (1997), and others (one disturbing feature of the expertise literature is how little the various researchers seem to know about each other). And AI predictions have none of the features that would predict good predictions; to use Shanteau's table on what predicts good or bad expert performance, with the features of AI timeline predictions marked in red:

Shanteau's Expert Predictions

Good performance	Poor performance
Static stimuli	Dynamic (changeable) stimuli
Decisions about things	Decisions about behavior
Experts agree on stimuli	Experts disagree on stimuli
More predictable problems	Less predictable problems
Some errors expected	Few errors expected
Repetitive tasks	Unique tasks
Feedback available	Feedback unavailable
Objective analysis available	Subjective analysis only
Problem decomposable	Problem not decomposable
Decision aids common	Decision aids rare

Feedback, the most important component in good expert prediction, is almost completely absent (immediate feedback, the crown jewel of expert competence, is completely absent). Experts disagree strongly on AI dates and the formats of AI, as we've seen. Finally, though better predictions are made by decomposing the problem, few predictors do so (possibly because of the magnitude of the problem).

A big warning here: it is easy to fall into the rhetorical trap of thinking "if it's uncertain, then we don't need to worry about it". That "argument", or variants of it, is often used to dismiss concerns about AI, even by very intelligent critics. Considering the argument deployed in other areas is enough to illustrate its weakness: "the enemy army could come down the right pass or the left one, we really don't know, so let's not worry about either", or "the virus you caught may or may not be infectious, I wouldn't worry about it if I were you." In fact, claiming that AIs are safe or impossible (which is what "don't worry" amounts to) is a very confident and specific prediction about the future of AI development. Hence, almost certainly wrong. Uncertainty is not safety's ally.

References

Armstrong, Stuart. Smarter than us: The rise of machine intelligence. MIRI, 2014.

Bostrom, Nick. Superintelligence: Paths, dangers, strategies. OUP Oxford, 2014.

Cooke, Roger M., and Louis HJ Goossens. "Expert judgement elicitation for risk assessments of critical infrastructures." Journal of risk research 7.6 (2004): 643–656.

Eden, Amnon H., Eric Steinhart, David Pearce, and James H. Moor. "Singularity hypotheses: an overview." In Singularity Hypotheses, pp. 1–12. Springer Berlin Heidelberg, 2012. Harvard

Kahneman, Daniel. Thinking, fast and slow. Macmillan, 2011.

Klein, Gary. "The recognition-primed decision (RPD) model: Looking back, looking forward." Naturalistic decision making (1997): 285–292.

Sandberg, Anders. "An overview of models of technological singularity." Roadmaps to AGI and the Future of AGI Workshop, Lugano, Switzerland, March. Vol. 8. 2010.

Shanteau, James, et al. "Performance-based assessment of expertise: How to decide if someone is an expert or not." European Journal of Operational Research 136.2 (2002): 253–263.

Tetlock, Philip. Expert political judgment: How good is it? How can we know?. Princeton University Press, 2005.

Part I
Risks of, and Responses to, the Journey to the Singularity

Chapter 2
Risks of the Journey to the Singularity

Kaj Sotala and Roman Yampolskiy

2.1 Introduction[1]

Many have argued that in the next twenty to one hundred years we will create artificial general intelligences [AGIs] (Baum et al. 2011; Sandberg and Bostrom 2011; Müller and Bostrom 2014).[2] Unlike current "narrow" AI systems, AGIs would perform at or above the human level not merely in particular domains (e.g., chess or arithmetic), but in a wide variety of domains, including novel ones.[3] They would have a robust understanding of natural language and be capable of general problem solving.

The creation of AGI could pose challenges and risks of varied severity for society, such as the possibility of AGIs outcompeting humans in the job market (Brynjolfsson and McAfee 2011). This article, however, focuses on the suggestion

---

[1]This chapter is based on three earlier publications (Sotala and Yampolskiy 2015; Sotala and Yampolskiy 2013; Yampolskiy 2013).

[2]Unlike the term "human-level AI," the term "Artificial General Intelligence" does not necessarily presume that the intelligence will be human-like.

[3]For this paper, we use a binary distinction between narrow AI and AGI. This is merely for the sake of simplicity we do not assume the actual difference between the two categories to necessarily be so clean-cut.

K. Sotala
Foundational Research Institute, Basel, Switzerland
e-mail: kaj.sotala@foundational-research.org

R. Yampolskiy (✉)
University of Louisville, Louisville, USA
e-mail: roman.yampolskiy@louisville.edu

V. Callaghan et al. (eds.), *The Technological Singularity*,
The Frontiers Collection, DOI 10.1007/978-3-662-54033-6_2

that AGIs may come to act in ways not intended by their creators, and in this way pose a *catastrophic* (Bostrom and Ćirković 2008) or even an *existential* (Bostrom 2002) risk to humanity.[4]

2.2 Catastrophic AGI Risk

We begin with a brief sketch of the argument that AGI poses a catastrophic risk to humanity. At least two separate lines of argument seem to support this conclusion. This argument will be further elaborated on in the following sections.

First, AI has already made it possible to automate many jobs (Brynjolfsson and McAfee 2011), and AGIs, when they are created, should be capable of performing *most* jobs better than humans (Hanson 2008; Bostrom 2014). As humanity grows increasingly reliant on AGIs, these AGIs will begin to wield more and more influence and power. Even if AGIs initially function as subservient tools, an increasing number of decisions will be made by autonomous AGIs rather than by humans. Over time it would become ever more difficult to replace the AGIs, even if they no longer remained subservient.

Second, there may be a sudden discontinuity in which AGIs rapidly become far more numerous or intelligent (Good 1965; Chalmers 2010; Bostrom 2014). This could happen due to (1) a conceptual breakthrough which makes it easier to run AGIs using far less hardware, (2) AGIs using fast computing hardware to develop ever-faster hardware, or (3) AGIs crossing a threshold in intelligence that allows them to carry out increasingly fast software self-improvement. Even if the AGIs were expensive to develop at first, they could be cheaply copied and could thus spread quickly once created.

Once they become powerful enough, AGIs might be a threat to humanity even if they are not actively malevolent or hostile. Mere indifference to human values—including human survival—could be sufficient for AGIs to pose an existential threat (Yudkowsky 2008a, 2011; Omohundro 2007, 2008; Bostrom 2014).

We will now lay out the above reasoning in more detail.

2.2.1 Most Tasks Will Be Automated

Ever since the Industrial Revolution, society has become increasingly automated. Brynjolfsson and McAfee (2011) argue that the current high unemployment rate in the United States is partially due to rapid advances in information technology,

[4]A catastrophic risk is something that might inflict serious damage to human well-being on a global scale and cause ten million or more fatalities (Bostrom and Ćirković 2008). An existential risk is one that threatens human extinction (Bostrom 2002). Many writers argue that AGI might be a risk of such magnitude (Butler 1863; Wiener 1960; Good 1965; Vinge 1993; Joy 2000; Yudkowsky 2008a; Bostrom 2014).

which has made it possible to replace human workers with computers faster than human workers can be trained in jobs that computers cannot yet perform. Vending machines are replacing shop attendants, automated discovery programs which locate relevant legal documents are replacing lawyers and legal aides, and automated virtual assistants are replacing customer service representatives.

Labor is becoming automated for reasons of cost, efficiency, and quality. Once a machine becomes capable of performing a task as well as (or almost as well as) a human, the cost of purchasing and maintaining it may be less than the cost of having a salaried human perform the same task. In many cases, machines are also capable of doing the same job faster, for longer periods, and with fewer errors. In addition to replacing workers entirely, machines may also take over aspects of jobs that were once the sole domain of highly trained professionals, making the job easier to perform by less-skilled employees (Whitby 1996).

If workers can be affordably replaced by developing more sophisticated AI, there is a strong economic incentive to do so. This is already happening with narrow AI, which often requires major modifications or even a complete redesign in order to be adapted for new tasks. "A Roadmap for US Robotics" (Hollerbach et al. 2009) calls for major investments into automation, citing the potential for considerable improvements in the fields of manufacturing, logistics, health care, and services. Similarly, the US Air Force Chief Scientist's (Dahm 2010) "Technology Horizons" report mentions "increased use of autonomy and autonomous systems" as a key area of research to focus on in the next decade, and also notes that reducing the need for manpower provides the greatest potential for cutting costs. In 2000, the US Congress instructed the armed forces to have one third of their deep strike force aircraft be unmanned by 2010, and one third of their ground combat vehicles be unmanned by 2015 (Congress 2000).

To the extent that an AGI could learn to do many kinds of tasks—or even *any* kind of task—without needing an extensive re-engineering effort, the AGI could make the replacement of humans by machines much cheaper and more profitable. As more tasks become automated, the bottlenecks for further automation will require adaptability and flexibility that narrow-AI systems are incapable of. These will then make up an increasing portion of the economy, further strengthening the incentive to develop AGI.

Increasingly sophisticated AI may eventually lead to AGI, possibly within the next several decades (Baum et al. 2011; Müller and Bostrom 2014). Eventually it will make economic sense to automate all or nearly all jobs (Hanson 2008; Hall 2008). As AGIs will possess many advantages over humans (Sotala 2012; Muehlhauser and Salamon 2012a, b; Bostrom 2014), a greater and greater proportion of the workforce will consist of intelligent machines.

2.2.2 AGIs Might Harm Humans

AGIs might bestow overwhelming military, economic, or political power on the groups that control them (Bostrom 2002, 2014). For example, automation could lead to an ever-increasing transfer of wealth and power to the owners of the AGIs (Brynjolfsson and McAfee 2011). AGIs could also be used to develop advanced weapons and plans for military operations or political takeovers (Bostrom 2002). Some of these scenarios could lead to catastrophic risks, depending on the capabilities of the AGIs and other factors.

Our focus is on the risk from the possibility that AGIs could behave in unexpected and harmful ways, even if the intentions of their owners were benign. Even modern-day narrow-AI systems are becoming autonomous and powerful enough that they sometimes take unanticipated and harmful actions before a human supervisor has a chance to react. To take one example, rapid automated trading was found to have contributed to the 2010 stock market "Flash Crash" (CFTC and SEC 2010).[5] Autonomous systems may also cause people difficulties in more mundane situations, such as when a credit card is automatically flagged as possibly stolen due to an unusual usage pattern (Allen et al. 2006), or when automatic defense systems malfunction and cause deaths (Shachtman 2007).

As machines become more autonomous, humans will have fewer opportunities to intervene in time and will be forced to rely on machines making good choices. This has prompted the creation of the field of "machine ethics" (Wallach and Allen 2009; Allen et al. 2006; Anderson and Anderson 2011), concerned with creating AI systems designed to make appropriate moral choices. Compared to narrow-AI systems, AGIs will be even more autonomous and capable, and will thus require even more robust solutions for governing their behavior.[6]

If some AGIs were both powerful and indifferent to human values, the consequences could be disastrous. At one extreme, powerful AGIs indifferent to human survival could bring about human extinction. As Yudkowsky (2008a) writes, "The AI does not hate you, nor does it love you, but you are made out of atoms which it can use for something else."

---

[5]On the less serious front, see http://www.michaeleisen.org/blog/?p=358 for an amusing example of automated trading going awry.

[6]In practice, there have been two separate communities doing research on automated moral decision-making (Muehlhauser and Helm 2012a, b; Allen and Wallach 2012; Shulman et al. 2009). The "AI risk" community has concentrated specifically on advanced AGIs (e.g. Yudkowsky 2008a; Bostrom 2014), while the "machine ethics" community typically has concentrated on more immediate applications for current-day AI (e.g. Wallach et al. 2008; Anderson and Anderson 2011). In this chapter, we have cited the machine ethics literature only where it seemed relevant, leaving out papers that seemed to be too focused on narrow-AI systems for our purposes. In particular, we have left out most discussions of military machine ethics (Arkin 2009), which focus primarily on the constrained special case of creating systems that are safe for battlefield usage.

Omohundro (2007, 2008) and Bostrom (2012) argue that standard microeconomic theory prescribes particular instrumental behaviors which are useful for the achievement of almost any set of goals. Furthermore, any agents which do not follow certain axioms of rational behavior will possess vulnerabilities which some other agent may exploit to their own benefit. Thus AGIs which understand these principles and wish to act efficiently will modify themselves so that their behavior more closely resembles rational economic behavior (Omohundro 2012). Extra resources are useful in the pursuit of nearly any set of goals, and self-preservation behaviors will increase the probability that the agent can continue to further its goals. AGI systems which follow rational economic theory will then exhibit tendencies toward behaviors such as self-replicating, breaking into other machines, and acquiring resources without regard for anyone else's safety. They will also attempt to improve themselves in order to more effectively achieve these and other goals, which could lead to rapid improvement even if the designers did not intend the agent to self-improve.

Even AGIs that were explicitly designed to behave ethically might end up acting at cross-purposes to humanity, because it is difficult to precisely capture the complexity of human values in machine goal systems (Yudkowsky 2011; Muehlhauser and Helm 2012a, b; Bostrom 2014).

Muehlhauser and Helm (2012a, b) caution that moral philosophy has found no satisfactory formalization of human values. All moral theories proposed so far would lead to undesirable consequences if implemented by superintelligent machines. For example, a machine programmed to maximize the satisfaction of human (or sentient) preferences might simply modify people's brains to give them desires that are maximally easy to satisfy.

Intuitively, one might say that current moral theories are all *too simple*—even if they seem correct at first glance, they do not actually take into account all the things that we value, and this leads to a catastrophic outcome. This could be referred to as the *complexity of value thesis*. Recent psychological and neuroscientific experiments confirm that human values are highly complex (Muehlhauser and Helm 2012a, b), that the pursuit of pleasure is not the only human value, and that humans are often unaware of their own values.

Still, perhaps powerful AGIs would have desirable consequences so long as they were programmed to respect *most* human values. If so, then our inability to perfectly specify human values in AGI designs need not pose a catastrophic risk. Different cultures and generations have historically had very different values from each other, and it seems likely that over time our values would become considerably different from current-day ones. It could be enough to maintain some small set of core values, though what exactly would constitute a core value is unclear. For example, different people may disagree over whether freedom or well-being is a more important value.

Yudkowsky (2011) argues that, due to the fragility of value, the basic problem remains. He argues that, even if an AGI implemented *most* human values, the outcome might still be unacceptable. For example, an AGI which failed to incorporate the value of novelty could create a solar system filled with countless minds experiencing one highly optimal and satisfying experience over and over again, never doing or feeling anything else (Yudkowsky 2009).[7]

In this paper, we will frequently refer to the problem of "AGI safety" or "safe AGI," by which we mean the problem of ensuring that AGIs respect human values, or perhaps some extrapolation or idealization of human values. We do not seek to imply that current human values would be the best possible ones, that AGIs could not help us in developing our values further, or that the values of other sentient beings would be irrelevant. Rather, by "human values" we refer to the kinds of basic values that nearly all humans would agree upon, such as that AGIs forcibly reprogramming people's brains, or destroying humanity, would be a bad outcome. In cases where proposals related to AGI risk might change human values in some major but not as obviously catastrophic way, we will mention the possibility of these changes but remain agnostic on whether they are desirable or undesirable.

We conclude this section with one frequently forgotten point in order to avoid catastrophic risks or worse, it is not enough to ensure that only some AGIs are safe. Proposals which seek to solve the issue of catastrophic AGI risk need to also provide some mechanism for ensuring that *most* (or perhaps even "nearly all") AGIs are either created safe or prevented from doing considerable harm.

2.2.3 AGIs May Become Powerful Quickly

There are several reasons why AGIs may quickly come to wield unprecedented power in society. "Wielding power" may mean having direct decision-making power, or it may mean carrying out human decisions in a way that makes the decision maker reliant on the AGI. For example, in a corporate context an AGI could be acting as the executive of the company, or it could be carrying out countless low-level tasks which the corporation needs to perform as part of its daily operations.

Bugaj and Goertzel (2007) consider three kinds of AGI scenarios: capped intelligence, soft takeoff, and hard takeoff. In a *capped intelligence* scenario, all AGIs are prevented from exceeding a predetermined level of intelligence and remain at a level roughly comparable with humans. In a *soft takeoff* scenario, AGIs become far more powerful than humans, but on a timescale which permits ongoing human interaction during the ascent. Time is not of the essence, and learning

[7]Miller (2012) similarly notes that, despite a common belief to the contrary, it is impossible to write laws in a manner that would match our stated moral principles without a judge needing to use a large amount of implicit common-sense knowledge to correctly interpret them. "Laws shouldn't always be interpreted literally because legislators can't anticipate all possible contingencies. Also, humans' intuitive feel for what constitutes murder goes beyond anything we can commit to paper. The same applies to friendliness." (Miller 2012).

proceeds at a relatively human-like pace. In a *hard takeoff* scenario, an AGI will undergo an extraordinarily fast increase in power, taking effective control of the world within a few years or less.[8] In this scenario, there is little time for error correction or a gradual tuning of the AGI's goals.

The viability of many proposed approaches depends on the hardness of a takeoff. The more time there is to react and adapt to developing AGIs, the easier it is to control them. A soft takeoff might allow for an approach of incremental machine ethics (Powers 2011), which would not require us to have a complete philosophical theory of ethics and values, but would rather allow us to solve problems in a gradual manner. A soft takeoff might however present its own problems, such as there being a larger number of AGIs distributed throughout the economy, making it harder to contain an eventual takeoff.

Hard takeoff scenarios can be roughly divided into those involving the quantity of hardware (the *hardware overhang* scenario), the quality of hardware (the *speed explosion* scenario), and the quality of software (the *intelligence explosion* scenario). Although we discuss them separately, it seems plausible that several of them could happen simultaneously and feed into each other.[9]

2.2.3.1 Hardware Overhang

Hardware progress may outpace AGI software progress. Contemporary supercomputers already rival or even exceed some estimates of the computational capacity of the human brain, while no software seems to have both the brain's general learning capacity and its scalability.[10]

If such trends continue, then by the time the software for AGI is invented there may be a *computing overhang*—an abundance of cheap hardware available for

---

[8]Bugaj and Goertzel defined hard takeoff to refer to a period of months or less. We have chosen a somewhat longer time period, as even a few years might easily turn out to be too little time for society to properly react.

[9]Bostrom (2014, chap. 3) discusses three kinds of superintelligence. A speed superintelligence "can do all that a human intellect can do, but much faster". A collective superintelligence is "a system composed of large number of smaller intellects such that the system's overall performance across many very general domains vastly outstrips that of any current cognitive system". A quality superintelligence "is at least as fast as a human mind and vastly qualitatively smarter". These can be seen as roughly corresponding to the different kinds of hard takeoff scenarios. A speed explosion implies a speed superintelligence, an intelligence explosion a quality superintelligence, and a hardware overhang may lead to any combination of speed, collective, and quality superintelligence.

[10]Bostrom (1998) estimates that the effective computing capacity of the human brain might be somewhere around 10^{17} operations per second (OPS), and Moravec (1998) estimates it at 10^{14} OPS. As of June 2016, the fastest supercomputer in the world had achieved a top capacity of 10^{16} floating-point operations per second (FLOPS) and the five-hundredth fastest a top capacity of 10^{14} FLOPS (Top500 2016). Note however that OPS and FLOPS are not directly comparable and there is no reliable way of interconverting the two. Sandberg and Bostrom (2008) estimate that OPS and FLOPS grow at a roughly comparable rate.

running thousands or millions of AGIs, possibly with a speed of thought much faster than that of humans (Yudkowsky 2008b; Shulman and Sandberg 2010, Sotala 2012).

As increasingly sophisticated AGI software becomes available, it would be possible to rapidly copy improvements to millions of servers, each new version being capable of doing more kinds of work or being run with less hardware. Thus, the AGI software could replace an increasingly large fraction of the workforce.[11] The need for AGI systems to be trained for some jobs would slow the rate of adoption, but powerful computers could allow for fast training. If AGIs end up doing the vast majority of work in society, humans could become dependent on them.

AGIs could also plausibly take control of Internet-connected machines in order to harness their computing power (Sotala 2012); Internet-connected machines are regularly compromised.[12]

2.2.3.2 Speed Explosion

Another possibility is a *speed explosion* (Solomonoff 1985; Yudkowsky 1996; Chalmers 2010), in which intelligent machines design increasingly faster machines. A hardware overhang might contribute to a speed explosion, but is not required for it. An AGI running at the pace of a human could develop a second generation of hardware on which it could run at a rate faster than human thought. It would then require a shorter time to develop a third generation of hardware, allowing it to run faster than on the previous generation, and so on. At some point, the process would hit physical limits and stop, but by that time AGIs might come to accomplish most tasks at far faster rates than humans, thereby achieving dominance. (In principle, the same process could also be achieved via improved software.)

---

[11]The speed that would allow AGIs to take over most jobs would depend on the cost of the hardware and the granularity of the software upgrades. A series of upgrades over an extended period, each producing a 1% improvement, would lead to a more gradual transition than a single upgrade that brought the software from the capability level of a chimpanzee to a rough human equivalence. Note also that several companies, including Amazon and Google, offer vast amounts of computing power for rent on an hourly basis. An AGI that acquired money and then invested all of it in renting a large amount of computing resources for a brief period could temporarily achieve a much larger boost than its budget would otherwise suggest.

[12]Botnets are networks of computers that have been compromised by outside attackers and are used for illegitimate purposes. Rajab et al. (2007) review several studies which estimate the sizes of the largest botnets as being between a few thousand to 350,000 bots. Modern-day malware could theoretically infect any susceptible Internet-connected machine within tens of seconds of its initial release (Staniford et al. 2002). The Slammer worm successfully infected more than 90% of vulnerable hosts within ten minutes, and had infected at least 75,000 machines by the thirty-minute mark (Moore et al. 2003). The previous record holder in speed, the Code Red worm, took fourteen hours to infect more than 359,000 machines (Moore et al. 2002).

The extent to which the AGI needs humans in order to produce better hardware will limit the pace of the speed explosion, so a rapid speed explosion requires the ability to automate a large proportion of the hardware manufacturing process. However, this kind of automation may already be achieved by the time that AGI is developed.[13]

2.2.3.3 Intelligence Explosion

Third, there could be an *intelligence explosion*, in which one AGI figures out how to create a qualitatively smarter AGI, and that AGI uses its increased intelligence to create still more intelligent AGIs, and so on,[14] such that the intelligence of humankind is quickly left far behind and the machines achieve dominance (Good 1965; Chalmers 2010; Muehlhauser and Salamon 2012a, b; Loosemore and Goertzel 2012; Bostrom 2014).

Yudkowsky (2008a, b) argues that an intelligence explosion is likely. So far, natural selection has been improving human intelligence, and human intelligence has to some extent been able to improve itself. However, the core process by which natural selection improves humanity has been essentially unchanged, and humans have been unable to deeply affect the cognitive algorithms which produce their own intelligence. Yudkowsky suggests that if a mind became capable of directly editing itself, this could spark a rapid increase in intelligence, as the actual process causing increases in intelligence could itself be improved upon. (This requires that there exist powerful improvements which, when implemented, considerably increase the rate at which such minds can improve themselves.)

Hall (2008) argues that, based on standard economic considerations, it would not make sense for an AGI to focus its resources on solitary self-improvement. Rather, in order not to be left behind by society at large, it should focus its resources on doing the things that it is good at and trade for the things it is not good at. However, once there exists a community of AGIs that can trade with one another, this community could collectively undergo rapid improvement and leave humans behind.

---

[13]Loosemore and Goertzel (2012) also suggest that current companies carrying out research and development are more constrained by a lack of capable researchers than by the ability to carry out physical experiments.

[14]Most accounts of this scenario do not give exact definitions for "intelligence" or explain what a "superintelligent" AGI would be like, instead using informal characterizations such as "a machine that can surpass the intellectual activities of any man however clever" (Good 1965) or "an intellect that is much smarter than the best human brains in practically every field, including scientific creativity, general wisdom and social skills" (Bostrom 1998). Yudkowsky (2008a) defines intelligence in relation to "optimization power," the ability to reliably hit small targets in large search spaces, such as by finding the a priori exceedingly unlikely organization of atoms which makes up a car. A more mathematical definition of machine intelligence is offered by Legg and Hutter (2007). Sotala (2012) discusses some of the functional routes to actually achieving superintelligence.

A number of formal growth models have been developed which are relevant to predicting the speed of a takeoff; an overview of these can be found in Sandberg (2010). Many of them suggest rapid growth. For instance, Hanson (1998) suggests that AGI might lead to the economy doubling in months rather than years. However, Hanson is skeptical about whether this would prove a major risk to humanity, and considers it mainly an economic transition similar to the Industrial Revolution.

To some extent, the soft/hard takeoff distinction may be a false dichotomy. A takeoff may be soft for a while, and then become hard. Two of the main factors influencing the speed of a takeoff are the pace at which computing hardware is developed and the ease of modifying minds (Sotala 2012). This allows for scenarios in which AGI is developed and there seems to be a soft takeoff for, say, the initial ten years, causing a false sense of security until a breakthrough in hardware development causes a hard takeoff.

Another factor that might cause a false sense of security is the possibility that AGIs can be developed by a combination of insights from humans and AGIs themselves. As AGIs become more intelligent and it becomes possible to automate portions of the development effort, those parts accelerate and the parts requiring human effort become bottlenecks. Reducing the amount of human insight required could dramatically accelerate the speed of improvement. Halving the amount of human involvement required might at most double the speed of development, possibly giving an impression of relative safety, but going from 50% human insight required to 1% human insight required could cause the development to become ninety-nine times faster.[15]

From a safety viewpoint, the conservative assumption is to presume the worst (Yudkowsky 2001). Yudkowsky argues that the worst outcome would be a hard takeoff, as it would give us the least time to prepare and correct errors. On the other hand, it can also be argued that a soft takeoff would be just as bad, as it would allow the creation of multiple competing AGIs, allowing the AGIs that were the least burdened with goals such as "respect human values" to prevail. We would ideally like a solution, or a combination of solutions, which would work effectively for both a soft and a hard takeoff.

References

Allen, Colin, and Wendell Wallach. 2012. "Moral Machines: Contradiction in Terms or Abdication of Human Responsibility." In Lin, Abney, and Bekey 2012, 55–68.

Allen, Colin, Wendell Wallach, and Iva Smit. 2006. "Why Machine Ethics?" IEEE Intelligent Systems 21 (4): 12–17. doi:10.1109/MIS.2006.83.

Amdahl, Gene M. 1967. "Validity of the Single Processor Approach to Achieving Large Scale Computing Capabilities." In Proceedings of the April 18–20, 1967, Spring Joint Computer

[15]The relationship in question is similar to that described by Amdahl's (1967) law.

Conference—AFIPS '67 (Spring), 483–485. New York: ACM Press. doi:10.1145/1465482.1465560.

Anderson, Michael, and Susan Leigh Anderson, eds. 2011. Machine Ethics. New York: Cambridge University Press.

Arkin, Ronald C. 2009. Governing Lethal Behavior in Autonomous Robots. Boca Raton, FL: CRC Press.

Baum, Seth D., Ben Goertzel, and Ted G. Goertzel. 2011. "How Long Until Human-Level AI? Results from an Expert Assessment." Technological Forecasting and Social Change 78 (1): 185–195. doi:10.1016/j.techfore.2010.09.006.

Bostrom, Nick. 1998. "How Long Before Superintelligence?" International Journal of Futures Studies 2.

Bostrom, Nick. 2002. "Existential Risks: Analyzing Human Extinction Scenarios and Related Hazards." Journal of Evolution and Technology 9. http://www.jetpress.org/volume9/risks.html.

Bostrom, Nick. 2012. "The Superintelligent Will: Motivation and Instrumental Rationality in Advanced Artificial Agents." In "Theory and Philosophy of AI," edited by Vincent C. Müller. Special issue, Minds and Machines 22 (2): 71–85. doi:10.1007/s11023-012-9281-3.

Bostrom, Nick. 2014. Superintelligence: Paths, dangers, strategies. Oxford University Press.

Bostrom, Nick, and Milan M. Ćirković. 2008. "Introduction." In Bostrom, Nick, and Milan M. Ćirković, eds. Global Catastrophic Risks. New York: Oxford University Press., 1–30.

Brynjolfsson, Erik, and Andrew McAfee. 2011. Race Against The Machine: How the Digital Revolution is Accelerating Innovation, Driving Productivity, and Irreversibly Transforming Employment and the Economy. Lexington, MA: Digital Frontier. Kindle edition.

Bugaj, Stephan Vladimir, and Ben Goertzel. 2007. "Five Ethical Imperatives and Their Implications for Human-AGI Interaction." Dynamical Psychology. http://goertzel.org/dynapsyc/2007/Five_Ethical_Imperatives_svbedit.htm.

Butler, Samuel [Cellarius, pseud.]. 1863. "Darwin Among the Machines." Christchurch Press, June 13. http://www.nzetc.org/tm/scholarly/tei-ButFir-t1-g1-t1-g1-t4-body.html.

CFTC & SEC (Commodity Futures Trading Commission and Securities & Exchange Commission). 2010. Findings Regarding the Market Events of May 6, 2010: Report of the Staffs of the CFTC and SEC to the Joint Advisory Committee on Emerging Regulatory Issues. Washington, DC. http://www.sec.gov/news/studies/2010/marketevents-report.pdf.

Chalmers, David John. 2010. "The Singularity: A Philosophical Analysis." Journal of Consciousness Studies 17 (9–10): 7–65. http://www.ingentaconnect.com/content/imp/jcs/2010/00000017/f0020009/art00001.

Congress, US. 2000. National Defense Authorization, Fiscal Year 2001, Pub. L. No. 106–398, 114 Stat. 1654.

Dahm, Werner J. A. 2010. Technology Horizons: A Vision for Air Force Science & Technology During 2010-2030. AF/ST-TR-10-01-PR. Washington, DC: USAF. http://www.au.af.mil/au/awc/awcgate/af/tech_horizons_vol-1_may2010.pdf.

Good, Irving John. 1965. "Speculations Concerning the First Ultraintelligent Machine." In Advances in Computers, edited by Franz L. Alt and Morris Rubinoff, 31–88. Vol. 6. New York: Academic Press. doi:10.1016/S0065-2458(08)60418-0.

Hall, John Storrs 2008. "Engineering Utopia." In Wang, Goertzel, and Franklin 2008, 460–467.

Hanson, Robin. 1998. "Economic Growth Given Machine Intelligence." Unpublished manuscript. Accessed May 15, 2013. http://hanson.gmu.edu/aigrow.pdf.

Hanson, Robin. 2008. "Economics of the Singularity." IEEE Spectrum 45 (6): 45–50. doi:10.1109/MSPEC.2008.4531461.

Hollerbach, John M., Matthew T. Mason, and Henrik I. Christensen. 2009. A Roadmap for US Robotics: From Internet to Robotics. Snobird, UT: Computing Community Consortium. http://www.usrobotics.us/reports/CCC%20Report.pdf.

Joy, Bill. 2000. "Why the Future Doesn't Need Us." Wired, April. http://www.wired.com/wired/archive/8.04/joy.html.

Legg, Shane, and Marcus Hutter. 2007. "A Collection of Definitions of Intelligence." In Advances in Artificial General Intelligence: Concepts, Architectures and Algorithms—Proceedings of the

AGI Workshop 2006, edited by Ben Goertzel and Pei Wang, 17–24. Frontiers in Artificial Intelligence and Applications 157. Amsterdam: IOS.

Loosemore, Richard, and Ben Goertzel. 2012. "Why an Intelligence Explosion is Probable." In Eden, Amnon, Johnny Søraker, James H. Moor, and Eric Steinhart, eds. Singularity Hypotheses: A Scientific and Philosophical Assessment. The Frontiers Collection. Berlin: Springer.

Miller, James D. 2012. Singularity Rising: Surviving and Thriving in a Smarter, Richer, and More Dangerous World. Dallas, TX: BenBella Books.

Moore, David, Vern Paxson, Stefan Savage, Colleen Shannon, Stuart Staniford, and Nicholas Weaver. 2003. "Inside the Slammer Worm." IEEE Security & Privacy Magazine 1 (4): 33–39. doi:10.1109/MSECP.2003.1219056.

Moore, David, Colleen Shannon, and Jeffery Brown. 2002. "Code-Red: A Case Study on the Spread and Victims of an Internet Worm." In Proceedings of the Second ACM SIGCOMM Workshop on Internet Measurement (IMW '02), 273–284. New York: ACM Press. doi:10.1145/637201.637244.

Moravec, Hans P. 1998. "When Will Computer Hardware Match the Human Brain?" Journal of Evolution and Technology 1. http://www.transhumanist.com/volume1/moravec.htm.

Muehlhauser, Luke, and Louie Helm. 2012. "The Singularity and Machine Ethics." In Eden, Amnon, Johnny Søraker, James H. Moor, and Eric Steinhart, eds. Singularity Hypotheses: A Scientific and Philosophical Assessment. The Frontiers Collection. Berlin: Springer.

Muehlhauser, Luke, and Anna Salamon. 2012. "Intelligence Explosion: Evidence and Import." In Eden, Amnon, Johnny Søraker, James H. Moor, and Eric Steinhart, eds. Singularity Hypotheses: A Scientific and Philosophical Assessment. The Frontiers Collection. Berlin: Springer.

Müller, V. C., and Bostrom, N. 2014. Future progress in artificial intelligence: A survey of expert opinion. Fundamental Issues of Artificial Intelligence.

Omohundro, Stephen M. 2007. "The Nature of Self-Improving Artificial Intelligence." Paper presented at Singularity Summit 2007, San Francisco, CA, September 8–9. http://selfawaresystems.com/2007/10/05/paper-on-the-nature-of-self-improving-artificial-intelligence/.

Omohundro, Stephen M. 2008. "The Basic AI Drives." In Wang, Goertzel, and Franklin 2008, 483–492.

Omohundro, Stephen M. 2012. "Rational Artificial Intelligence for the Greater Good." In Eden, Amnon, Johnny Søraker, James H. Moor, and Eric Steinhart, eds. Singularity Hypotheses: A Scientific and Philosophical Assessment. The Frontiers Collection. Berlin: Springer.

Powers, Thomas M. 2011. "Incremental Machine Ethics." IEEE Robotics & Automation Magazine 18 (1): 51–58. doi:10.1109/MRA.2010.940152.

Rajab, Moheeb Abu, Jay Zarfoss, Fabian Monrose, and Andreas Terzis. 2007. "My Botnet is Bigger than Yours (Maybe, Better than Yours): Why Size Estimates Remain Challenging." In Proceedings of 1st Workshop on Hot Topics in Understanding Botnets (HotBots '07).Berkeley, CA: USENIX. http://static.usenix.org/event/hotbots07/tech/full_papers/rajab/rajab.pdf.

Sandberg, Anders. 2010. "An Overview of Models of Technological Singularity." Paper presented at the Roadmaps to AGI and the Future of AGI Workshop, Lugano, Switzerland, March 8. http://agi-conf.org/2010/wp-content/uploads/2009/06/agi10singmodels2.pdf.

Sandberg, Anders, and Nick Bostrom. 2008. Whole Brain Emulation: A Roadmap. Technical Report, 2008-3. Future of Humanity Institute, University of Oxford. http://www.fhi.ox.ac.uk/wpcontent/uploads/brain-emulation-roadmap-report1.pdf.

Sandberg, Anders, and Nick Bostrom. 2011. Machine Intelligence Survey. Technical Report, 2011-1. Future of Humanity Institute, University of Oxford. http://www.fhi.ox.ac.uk/reports/2011-1.pdf.

Shachtman, Noah. 2007. "Robot Cannon Kills 9, Wounds 14." Wired, October 18. http://www.wired.com/dangerroom/2007/10/robot-cannon-ki/.

Shulman, Carl, and Anders Sandberg. 2010. "Implications of a Software-Limited Singularity." In Mainzer, Klaus, ed. ECAP10: VIII European Conference on Computing and Philosophy. Munich: Dr. Hut.

Shulman, Carl, Henrik Jonsson, and Nick Tarleton. 2009. "Machine Ethics and Superintelligence." In Reynolds, Carson, and Alvaro Cassinelli, eds. AP-CAP 2009: The Fifth Asia-Pacific Computing and Philosophy Conference, October 1st-2nd, University of Tokyo, Japan, Proceedings, 95–97.

Solomonoff, Ray J. 1985. "The Time Scale of Artificial Intelligence: Reflections on Social Effects." Human Systems Management 5:149–153.

Sotala, Kaj, and Roman V. Yampolskiy. 2013. Responses to catastrophic AGI risk: a survey. Technical report 2013-2. Berkeley, CA: Machine Intelligence Research Institute.

Sotala, Kaj, and Roman V. Yampolskiy. 2015. Responses to catastrophic AGI risk: a survey. Physica Scripta, 90(1), 018001.

Sotala, Kaj. 2012. "Advantages of Artificial Intelligences, Uploads, and Digital Minds." International Journal of Machine Consciousness 4 (1): 275–291. doi:10.1142/S1793843012400161.

Staniford, Stuart, Vern Paxson, and Nicholas Weaver. 2002. "How to 0wn the Internet in Your Spare Time." In Proceedings of the 11th USENIX Security Symposium, edited by Dan Boneh, 149–167. Berkeley, CA: USENIX. http://www.icir.org/vern/papers/cdc-usenix-sec02/.

Top500.org. 2016. Top500 list – June 2016. https://www.top500.org/list/2016/06/.

Vinge, Vernor. 1993. "The Coming Technological Singularity: How to Survive in the Post-Human Era." In Vision-21: Interdisciplinary Science and Engineering in the Era of Cyberspace, 11–22. NASA Conference Publication 10129. NASA Lewis Research Center. http://ntrs.nasa.gov/archive/nasa/casi.ntrs.nasa.gov/19940022855_1994022855.pdf.

Wallach, Wendell, and Colin Allen. 2009. Moral Machines: Teaching Robots Right from Wrong. New York: Oxford University Press. doi:10.1093/acprof:oso/9780195374049.001.0001.

Wallach, Wendell, Colin Allen, and Iva Smit. 2008. "Machine Morality: Bottom-Up and Top-Down Approaches for Modelling Human Moral Faculties." In "Ethics and Artificial Agents." Special issue, AI & Society 22 (4): 565–582. doi:10.1007/s00146-007-0099-0.

Whitby, Blay. 1996. Reflections on Artificial Intelligence: The Legal, Moral, and Ethical Dimensions. Exeter, UK: Intellect Books.

Wiener, Norbert. 1960. "Some Moral and Technical Consequences of Automation." Science 131 (3410): 1355–1358. http://www.jstor.org/stable/1705998.

Yampolskiy, Roman V. 2013. What to Do with the Singularity Paradox? Studies in Applied Philosophy, Epistemology and Rational Ethics vol 5, pp. 397–413. Springer Berlin Heidelberg.

Yudkowsky, Eliezer. 1996. "Staring into the Singularity." Unpublished manuscript. Last revised May 27, 2001. http://yudkowsky.net/obsolete/singularity.html.

Yudkowsky, Eliezer. 2001. Creating Friendly AI 1.0: The Analysis and Design of Benevolent Goal Architectures. The Singularity Institute, San Francisco, CA, June 15. http://intelligence.org/files/CFAI.pdf.

Yudkowsky, Eliezer. 2008a. "Artificial Intelligence as a Positive and Negative Factor in Global Risk." In Bostrom, Nick, and Milan M. Ćirković, eds. Global Catastrophic Risks. New York: Oxford University Press., 308–345.

Yudkowsky, Eliezer. 2008b. "Hard Takeoff." Less Wrong (blog), December 2. http://lesswrong.com/lw/wf/hard_takeoff/.

Yudkowsky, Eliezer. 2009. "Value is Fragile." Less Wrong (blog), January 29. http://lesswrong.com/lw/y3/value_is_fragile/.

Yudkowsky, Eliezer. 2011. Complex Value Systems are Required to Realize Valuable Futures. The Singularity Institute, San Francisco, CA. http://intelligence.org/files/ComplexValues.pdf.

Chapter 3
Responses to the Journey to the Singularity

Kaj Sotala and Roman Yampolskiy

3.1 Introduction

The notion of catastrophic AGI risk is not new, and this concern was expressed by early thinkers in the field. Hence, there have also been many proposals concerning what to do about it. The proposals we survey are neither exhaustive nor mutually exclusive: the best way of achieving a desirable outcome may involve pursuing several proposals simultaneously.

Section 3.2 briefly discusses some of the most recent developments in the field. Sections 3.3–3.5 survey three categories of proposals for dealing with AGI risk: societal proposals, proposals for external constraints on AGI behaviors, and proposals for creating AGIs that are safe due to their internal design. Although the main purpose of this paper is to provide a summary of existing work, we briefly provide commentary on the proposals in each major subsection of Sects. 3.3–3.5 and highlight some of the proposals we consider the most promising in Sect. 3.6. which are, regulation (Sect. 3.3.3), merging with machines (Sect. 3.3.4), AGI confinement (Sect. 3.4.1), Oracle AI (Sect. 3.5.1), and motivational weaknesses (Sect. 3.5.7).

In the long term, the most promising approaches seem to be value learning (Sect. 3.5.2.5) and human-like architectures (Sect. 3.5.3.4). Section 3.6 provides an extended discussion of the various merits and problems of these proposals.

Foundational Research Institute (this work was written while the author was at the Machine Intelligence Research Institute)

K. Sotala
Foundational Research Institute, Basel, Switzerland
e-mail: kaj.sotala@foundational-research.org

R. Yampolskiy (✉)
University of Louisville, 222 Eastern Parkway, Louisville, KY 40292, USA
e-mail: roman.yampolskiy@louisville.edu

V. Callaghan et al. (eds.), *The Technological Singularity*,
The Frontiers Collection, DOI 10.1007/978-3-662-54033-6_3

3.2 *Post-Superintelligence* Responses

This chapter is based on an earlier paper (Sotala and Yampolskiy 2015), which was the formally published version of a previous technical report (Sotala and Yampolskiy 2013). The tech report, in turn, was a greatly expanded version of an earlier conference paper (Yampolskiy 2013). Since the writing of the original papers, the topic of catastrophic AGI risk has attracted considerable attention both in academia and the popular press, much of it due to the publication of the book *Superintelligence* (Bostrom 2014).

We feel that it would not be appropriate to simply lump in all the new responses together with the old sections, as the debate has now become considerably more active and high-profile. In particular, numerous AI researchers have signed an open letter calling for more research into making sure that AI systems will be robust and beneficial rather than just capable (Future of Life Institute 2015). The open letter included a list of suggested research directions (Russell et al. 2015), including ones specifically aimed at dealing with the risks from AGI. The research directions document draws on a number of sources, including an ambitious research agenda recently published by the Machine Intelligence Research Institute (see Chap. 5). Soon after the publication of the open letter, Elon Musk donated 10 million dollars for the purpose of furthering research into safe and beneficial AI and AGI.

At the same time, several prominent researchers have also expressed the feeling that the risks from AGI are overhyped, and that there is a danger of the general public taking them too seriously at this stage. This position has been expressed in interviews of researchers such as Professor Andrew Ng (Madrigal 2015) and Facebook AI director Yann LeCun (Gomes 2015), who emphasize that current-day technology is still a long way from AGI. Even the more skeptical researchers tend to agree that the issue will eventually require some consideration, however (Alexander 2015).

3.3 Societal Proposals

Proposals can be divided into three general categories proposals for societal action, design proposals for external constraints on AGI behavior, and design recommendations for internal constraints on AGI behavior. In this section we briefly survey societal proposals. These include doing nothing, integrating AGIs with society, regulating research, merging with machines, and relinquishing research into AGI.

3.3.1 Do Nothing

3.3.1.1 AI Is Too Distant to Be Worth Our Attention

One response is that, although AGI is possible in principle, there is no reason to expect it in the near future. Typically, this response arises from the belief that, although there have been great strides in narrow AI, researchers are still very far from understanding how to build AGI. Distinguished computer scientists such as Gordon Bell and Gordon Moore, as well as cognitive scientists such as Douglas Hofstadter and Steven Pinker, have expressed the opinion that the advent of AGI is remote (IEEE Spectrum 2008). Davis (2012) reviews some of the ways in which computers are still far from human capabilities. Bringsjord and Bringsjord (2012) even claim that a belief in AGI this century is fideistic, appropriate within the realm of religion but not within science or engineering.

Some writers also actively criticize any discussion of AGI risk in the first place. The philosopher Alfred Nordmann (2007, 2008) holds the view that ethical concern is a scarce resource, not to be wasted on unlikely future scenarios such as AGI. Likewise, Dennett (2012) considers AGI risk an "imprudent pastime" because it distracts our attention from more immediate threats.

Others think that AGI is far off and not yet a major concern, but admit that it might be valuable to give the issue some attention. A presidential panel of the Association for the Advancement of Artificial Intelligence considering the long-term future of AI concluded that there was overall skepticism about AGI risk, but that additional research into the topic and related subjects would be valuable (Horvitz and Selman 2009). Posner (2004) writes that dedicated efforts for addressing the problem can wait, but that we should gather more information about the problem in the meanwhile.

Potential negative consequences of AGI are enormous, ranging from economic instability to human extinction. "Do nothing" could be a reasonable course of action if near-term AGI seemed extremely unlikely, if it seemed too early for any proposals to be effective in reducing risk, or if those proposals seemed too expensive to implement.

As a comparison, asteroid impact prevention is generally considered a topic worth studying, even though the probability of a civilization-threatening asteroid impact in the near future is not considered high. Napier (2008) discusses several ways of estimating the frequency of such impacts. Many models produce a rate of one civilization-threatening impact per five hundred thousand or more years, though some models suggest that rates of one such impact per hundred thousand years cannot be excluded.

An estimate of one impact per hundred thousand years would suggest less than a 0.1% chance of a civilization-threatening impact within the next hundred years. The probability of AGI being developed within the same period seems considerably higher (Müller and Bostrom 2014), and there is likewise a reasonable chance of a hard takeoff after it has been developed (Yudkowsky 2008, 2008b), suggesting that the topic is at the very least worth studying. Even without a hard takeoff, society is becoming increasingly automated, and even narrow AI is starting to require ethical guidelines (Wallach and Allen 2009).

We know neither which fields of science will be needed nor how much progress in them will be necessary for safe AGI. If much progress is needed and we believe effective progress to be possible this early on, it becomes reasonable to start studying the topic even before AGI is near. Muehlhauser and Helm (2012) suggest that, for one safe AGI approach alone (value learning, discussed further in Sect. 3.5.2.5), efforts by AGI researchers, economists, mathematicians, and philosophers may be needed. Safe AI may require the solutions for some of these problems to come well before AGI is developed.

3.3.1.2 Little Risk, no Action Needed

Some authors accept that a form of AGI will probably be developed but do not consider autonomous AGI to be a risk, or consider the possible negative consequences acceptable. Bryson and Kime (1998) argue that, although AGI will require us to consider ethical and social dangers, the dangers will be no worse than those of other technologies. Whitby (1996) writes that there has historically been no consistent trend of the most intelligent people acquiring the most authority, and that computers will augment humans rather than replace them. Whitby and Oliver (2000) further argue that AGIs will not have any particular motivation to act against us. Jenkins (2003) agrees with these points to the extent of saying that a machine will only act against humans if it is programmed to value itself over humans, although she does find AGI to be a real concern.

Another kind of "no action needed" response argues that AGI development will take a long time (Brooks 2008), implying that there will be plenty of time to deal with the issue later on. This can also be taken as an argument for later efforts being more effective, as they will be better tuned to AGI as it develops.

Others argue that superintelligence will not be possible at all.[1] McDermott (2012) points out that there are no good examples of algorithms which could be improved upon *indefinitely*. Deutsch (2011) argues that there will never be superintelligent AGIs, because human minds are already universal reasoners, and computers can at best speed up the experimental work that is required for testing and fine-tuning theories. He also suggests that even as the speed of technological development increases, so will our ability to deal with change. Anderson (2010) likewise suggests that the inherent unpredictability of the world will place upper limits on an entity's effective intelligence.

---

[1]The opposite argument is that superior intelligence will inevitably lead to more moral behavior. Some of the arguments related to this position are discussed in the context of evolutionary invariants (Sect. 3.5.3.1), although the authors advocating the use of evolutionary invariants do believe AGI risk to be worth our concern.

Heylighen (2012) argues that a single, stand-alone computer is exceedingly unlikely to become superintelligent, and that individual intelligences are always outmatched by the distributed intelligence found in social systems of many minds. Superintelligence will be achieved by building systems that integrate and improve the "Global Brain," the collective intelligence of everyone on Earth. Heylighen does acknowledge that this kind of a transition will pose its own challenges, but not of the kind usually evoked in discussions of AGI risk.

The idea of AGIs not having a motivation to act against humans is intuitively appealing, but there seem to be strong theoretical arguments against it. As mentioned earlier, Omohundro (2007, 2008) and Bostrom (2012) argue that self-replication and the acquisition of resources are useful in the pursuit of many different kinds of goals, and that many types of AI systems will therefore exhibit tendencies toward behaviors such as breaking into other machines, self-replicating, and acquiring resources without regard for anyone else's safety. The right design might make it possible to partially work around these behaviors (Shulman 2010a; Wang 2012), but they still need to be taken into account. Furthermore, we might not foresee all the complex interactions of different AGI mechanisms in the systems that we build, and they may end up with very different goals than the ones we intended (Yudkowsky 2008, 2011).

Can AGIs become superintelligent? First, we note that AGIs do not necessarily need to be much more intelligent than humans in order to be dangerous. AGIs already enjoy advantages such as the ability to rapidly expand their population by having themselves copied (Hanson 1994, 2008; Sotala 2012a), which may confer on them considerable economic and political influence even if they were not superintelligent. A better-than-human ability to coordinate their actions, which AGIs of a similar design could plausibly have (Sotala 2012), might then be enough to tilt the odds in their favor.

Another consideration is that AGIs do not necessarily need to be qualitatively more intelligent than humans in order to outperform humans. An AGI that merely thought twice as fast as any single human could still defeat him at intellectual tasks that had a time constraint, all else equal. Here an "intellectual" task should be interpreted broadly to refer not only to "book smarts" but to any task that animals cannot perform due to their mental limitations—including tasks involving social skills (Yudkowsky 2008). Straightforward improvements in computing power could provide AGIs with a considerable advantage in speed, which the AGI could then use to study and accumulate experiences that improved its skills.

As for Heylighen's (2012) Global Brain argument, there does not seem to be a reason to presume that powerful AGIs could not be geographically distributed, or that they couldn't seize control of much of the Internet. Even if individual minds were not very smart and needed a society to make progress, for minds that are capable of copying themselves and communicating perfectly with each other, individual instances of the mind might be better understood as parts of a whole than as separate individuals. In general, the distinction between an individual and a community might not be meaningful for AGIs. If there were enough AGIs, they might be able to form a community sufficient to take control of the rest of the Earth.

Heylighen (2007) himself has argued that many of the features of the Internet are virtually identical to the mechanisms used by the human brain. If the AGI is not carefully controlled, it might end up in a position where it made up the majority of the "Global Brain" and could undertake actions which the remaining parts of the organism did not agree with.

3.3.1.3 Let Them Kill Us

Dietrich (2007) argues that humanity frequently harms other species, and that people have also evolved to hurt other people by engaging in behaviors such as child abuse, sexism, rape, and racism. Therefore, human extinction would not matter, as long as the machines implemented only the positive aspects of humanity.

De Garis (2005) suggests that AGIs destroying humanity might not matter. He writes that on a cosmic scale, with hundreds of billions of stars in our galaxy alone, the survival of the inhabitants of a single planet is irrelevant. As AGIs would be more intelligent than us in every way, it would be better if they replaced humanity.

AGIs being more intelligent and therefore more valuable than humans equates intelligence with value, but Bostrom (2004) suggests ways by which a civilization of highly intelligent entities might lack things which we thought to have value. For example, such entities might not be conscious in the first place. Alternatively, there are many things which we consider valuable for their own sake, such as humor, love, game-playing, art, sex, dancing, social conversation, philosophy, literature, scientific discovery, food and drink, friendship, parenting, and sport. We value these due to the fact that we have dispositions and preferences which have been evolutionarily adaptive in the past, but for a future civilization few or none of them might be, creating a world with very little of value. Bostrom (2012) proposes an orthogonality thesis, by which an artificial intelligence can have any combination of intelligence level and goal, including goals that humans would intuitively deem to be of no value.

3.3.1.4 "Do Nothing" Proposals—Our View

As discussed above, completely ignoring the possibility of AGI risk at this stage would seem to require a confident belief in at least one of the following propositions

1. AGI is very remote.
2. There is no major risk from AGI even if it is created.
3. Very little effective work can be done at this stage.
4. AGIs destroying humanity would not matter.

In the beginning of this paper, we mentioned several experts who considered it plausible that AGI might be created in the next twenty to one hundred years; in this section we have covered experts who disagree.

In general, there is a great deal of disagreement among people who have made AGI predictions, and no clear consensus even among experts in the field of artificial intelligence. The lack of expert agreement suggests that expertise in the field does not contribute to an ability to make reliable predictions.[2] If the judgment of experts is not reliable, then, probably, neither is anyone else's. This suggests that it is unjustified to be highly certain of AGI being near, but also of it *not* being near. We thus consider it unreasonable to have a confident belief in the first proposition.

The second proposition also seems questionable. As discussed in the previous chapter, AGIs seem very likely to obtain great power, possibly very quickly. Furthermore, as also discussed in the previous chapter, the complexity and fragility of value theses imply that it could be very difficult to create AGIs which would not cause immense amounts of damage if they had enough power.

It also does not seem like it is too early to work on the problem as we summarize in Sect. 3.6, there seem to be a number of promising research directions which can already be pursued. We also agree with Yudkowsky (2008), who points out that research on the philosophical and technical requirements of safe AGI might show that broad classes of possible AGI architectures are fundamentally unsafe, suggesting that such architectures should be avoided. If this is the case, it seems better to have that knowledge as early as possible, before there has been a great deal of investment into unsafe AGI designs.

In response to the suggestion that humanity being destroyed would not matter, we certainly agree that there is much to be improved in today's humanity, and that our future descendants might have very little resemblance to ourselves. Regardless, we think that much about today's humans is valuable and worth preserving, and that we should be able to preserve it without involving the death of present humans.

3.3.2 Integrate with Society

Integration proposals hold that AGI might be created in the next several decades, and that there are indeed risks involved. These proposals argue that the best way to deal with the problem is to make sure that our societal structures are equipped to handle AGIs once they are created.

---

[2]Armstrong and Sotala (2012) point out that many of the task properties which have been found to be conducive for developing reliable and useful expertise are missing in AGI timeline forecasting. In particular, one of the most important factors is whether experts get rapid (preferably immediate) feedback, while a timeline prediction that is set many decades in the future might have been entirely forgotten by the time that its correctness could be evaluated.

There has been some initial work toward integrating AGIs with existing legal and social frameworks, such as considering questions of their legal position and moral rights (Gunkel 2012).

3.3.2.1 Legal and Economic Controls

Hanson (2012) writes that the values of older and younger generations have often been in conflict with each other, and he compares this to a conflict between humans and AGIs. He believes that the best way to control AGI risk is to create a legal framework such that it is in the interest of both humans and AGIs to uphold it. Hanson (2009) suggests that if the best way for AGIs to get what they want is via mutually agreeable exchanges, then humans would need to care less about what the AGIs wanted. According to him, we should be primarily concerned with ensuring that the AGIs will be law-abiding enough to respect our property rights. Miller (2012) summarizes Hanson's argument, and the idea that humanity could be content with a small fraction of the world's overall wealth and let the AGIs have the rest. An analogy to this idea is that humans do not kill people who become old enough to no longer contribute to production, even though younger people could in principle join together and take the wealth of the older people. Instead, old people are allowed to keep their wealth even while in retirement. If things went well, AGIs might similarly allow humanity to "retire" and keep its accumulated wealth, even if humans were no longer otherwise useful for AGIs.

Hall (2007a) also says that we should ensure that the interactions between ourselves and machines are economic, "based on universal rules of property and reciprocity." Moravec (1999) likewise writes that governmental controls should be used to ensure that humans benefit from AGIs. Without government intervention, humans would be squeezed out of existence by more efficient robots, but taxation could be used to support human populations for a long time. He also recommends laws which would require any AGIs to incorporate programming that made them safe and subservient to human desires. Sandberg (2001) writes that relying only on legal and economic controls would be problematic, but that a strategy which also incorporated them in addition to other approaches would be more robust than a strategy which did not.

However, even if AGIs were integrated with human institutions, it does not guarantee that human values would survive. If humans were reduced to a position of negligible power, AGIs might not have any reason to keep us around.

Economic arguments, such as the principle of comparative advantage, are sometimes invoked to argue that AGI would find it more beneficial to trade with us than to do us harm. However, technological progress can drive the wages of workers below the level needed for survival, and there is already a possible threat of technological unemployment (Brynjolfsson and McAfee 2011). AGIs keeping humans around due to gains from trade implicitly presumes that they would not have the will or the opportunity to simply eliminate humans in order to replace them with a better trading partner, and then trade with the new partner instead.

Humans already eliminate species with low economic value in order to make room for more humans, such as when clearing a forest in order to build new homes. Clark (2007) uses the example of horses in Britain their population peaked in 1901, with 3.25 million horses doing work such as plowing fields, hauling wagons and carriages short distances, and carrying armies into battle. The internal combustion engine replaced so many of them that by 1924 there were fewer than two million. Clark writes

> There was always a wage at which all these horses could have remained employed. But that wage was so low that it did not pay for their feed, and it certainly did not pay enough to breed fresh generations of horses to replace them. Horses were thus an early casualty of industrialization (Clark 2007).

There are also ways to harm humans while still respecting their property rights, such as by manipulating them into making bad decisions, or selling them addictive substances. If AGIs were sufficiently smarter than humans, humans could be tricked into making a series of trades that respected their property rights but left them with negligible assets and caused considerable damage to their well-being.

A related issue is that AGIs might become more capable of changing our values than we are capable of changing AGI values. Mass media already convey values that have a negative impact on human well-being, such as idealization of rare body types, which causes dissatisfaction among people who do not have those kinds of bodies (Groesz et al. 2001; Agliata and Tantleff-Dunn 2004). AGIs with a deep understanding of human psychology could engineer the spread of values which shifted more power to them, regardless of their effect on human well-being.

Yet another problem is ensuring that the AGIs have indeed adopted the right values. Making intelligent beings adopt specific values is a difficult process which often fails. There could be an AGI with the wrong goals that would pretend to behave correctly in society throughout the whole socialization process. AGIs could conceivably preserve and conceal their goals far better than humans could.

Society does not know of any methods which would reliably instill our chosen values in *human* minds, despite a long history of trying to develop them. Our attempts to make AGIs adopt human values would be hampered by our lack of experience and understanding of the AGI's thought processes, with even tried-and-true methods for instilling positive values in humans possibly being ineffective. The limited success that we do have with humans is often backed up by various incentives as well as threats of punishment, both of which might fail in the case of an AGI developing to become vastly more powerful than us.

Additionally, the values which a being is likely to adopt, or is even capable of adopting, will depend on its mental architecture. We will demonstrate these claims with examples from humans, who are not blank slates on whom arbitrary values can be imposed with the right education. Although the challenge of instilling specific values in humans is very different from the challenge of instilling them in AGIs, our examples are meant to demonstrate the fact that the existing properties of a mind will affect the process of acquiring values. Just as it is difficult to make humans permanently adopt some kinds of values, the kind of mental architecture that an AGI has will affect its inclination to adopt various values.

Psychopathy is a risk factor for violence, and psychopathic criminals are much more likely to reoffend than nonpsychopaths (Hare et al. 2000). Harris and Rice (2006) argue that therapy for psychopaths is ineffective and may even make them more dangerous, as they use their improved social skills to manipulate others more effectively. Furthermore, "cult brainwashing" is generally ineffective and most cult members will eventually leave (Anthony and Robbins 2004); and large-scale social engineering efforts often face widespread resistance, even in dictatorships with few scruples about which methods to use (Scott 1998, Chap. 6–7). Thus, while one can try to make humans adopt values, this will only work to the extent that the individuals in question are actually disposed toward adopting them.

3.3.2.2 Foster Positive Values

Kurzweil (2005), considering the possible effects of many future technologies, notes that AGI may be a catastrophic risk. He generally supports regulation and partial relinquishment of dangerous technologies, as well as research into their defensive applications. However, he believes that with AGI this may be insufficient and that, at the present time, it may be infeasible to develop strategies that would guarantee safe AGI. He argues that machine intelligences will be tightly integrated into our society and that, for the time being, the best chance of avoiding AGI risk is to foster positive values in our society. This will increase the likelihood that any AGIs that are created will reflect such positive values.

One possible way of achieving such a goal is moral enhancement (Douglas 2008), the use of technology to instill people with better motives. Persson and Savulescu (2008, 2012) argue that, as technology improves, we become more capable of damaging humanity, and that we need to carry out moral enhancement in order to lessen our destructive impulses.

3.3.2.3 "Integrate with Society" Proposals—Our View

Proposals to incorporate AGIs into society suffer from the issue that some AGIs may never adopt benevolent and cooperative values, no matter what the environment. Neither does the intelligence of the AGIs necessarily affect their values (Bostrom 2012). Sufficiently intelligent AGIs could certainly come to eventually understand human values, but humans can also come to understand others' values while continuing to disagree with them.

Thus, in order for these kinds of proposals to work, they need to incorporate strong enforcement mechanisms to keep non–safe AGIs in line and to prevent them from acquiring significant power. This requires an ability to create value-conforming AGIs in the first place, to implement the enforcement. Even a soft takeoff would eventually lead to AGIs wielding great power, so the enforcement could not be left to just humans or narrow AIs.[3] In practice, this means that integration proposals must be combined with some proposal for internal constraints which is capable of reliably creating value-conforming AGIs. Integration proposals also require there to be a soft takeoff in order to work, as having a small group of AGIs which rapidly acquired enough power to take control of the world would prevent any gradual integration schemes from working.

Therefore, because any effective integration strategy would require creating safe AGIs, and the right safe AGI design could lead to a positive outcome even if there were a hard takeoff, we believe that it is currently better to focus on proposals which are aimed at furthering the creation of safe AGIs.

3.3.3 Regulate Research

Integrating AGIs into society may require explicit regulation. Calls for regulation are often agnostic about long-term outcomes but nonetheless recommend caution as a reasonable approach. For example, Hibbard (2005b) calls for international regulation to ensure that AGIs will value the long-term well-being of humans, but does not go into much detail. Daley (2011) calls for a government panel for AGI issues. Hughes (2001) argues that AGI should be regulated using the same mechanisms as previous technologies, creating state agencies responsible for the task and fostering global cooperation in the regulation effort.

Current mainstream academic opinion does not consider AGI a serious threat (Horvitz and Selman 2009), so AGI regulation seems unlikely in the near future. On the other hand, many AI systems are becoming increasingly autonomous, and a number of authors are arguing that even narrow-AI applications should be equipped with an understanding of ethics (Wallach and Allen 2009). Currently there are calls to regulate AI in the form of high-frequency trading (Sobolewski 2012), and AI applications that have a major impact on society might become increasingly regulated. At the same time, legislation has a well-known tendency to lag behind technology, and regulating AI applications will probably not translate into regulating basic research into AGI.

3.3.3.1 Review Boards

Yampolskiy and Fox (2012) note that university research programs in the social and medical sciences are overseen by institutional review boards. They propose setting up analogous review boards to evaluate potential AGI research. Research that was found to be AGI related would be restricted with measures ranging from

[3]For proposals which suggest that humans could use technology to remain competitive with AGIs and thus prevent them from acquiring excessive amounts of power, see Sect. 3.4.

supervision and funding limits to partial or complete bans. At the same time, research focusing on safety measures would be encouraged.

Posner (2004, p. 221) suggests the enactment of a law which would require scientific research projects in dangerous areas to be reviewed by a federal catastrophic risks assessment board, and forbidden if the board found that the project would create an undue risk to human survival.

Wilson (2013) makes possibly the most detailed AGI regulation proposal so far, recommending a new international treaty where a body of experts would determine whether there was a "reasonable level of concern" about AGI or some other possibly dangerous research. States would be required to regulate research or even temporarily prohibit it once experts agreed upon there being such a level of concern. He also suggests a number of other safeguards built into the treaty, such as the creation of ethical oversight organizations for researchers, mechanisms for monitoring abuses of dangerous technologies, and an oversight mechanism for scientific publications.

3.3.3.2 Encourage Research into Safe AGI

In contrast, McGinnis (2010) argues that the government should not attempt to regulate AGI development. Rather, it should concentrate on providing funding for research projects intended to create safe AGI.

Goertzel and Pitt (2012) argue for an open-source approach to safe AGI development instead of regulation. Hibbard (2008) has likewise suggested developing AGI via open-source methods, but not as an alternative to regulation.

Legg (2009) proposes funding safe AGI research via an organization that takes a venture capitalist approach to funding research teams, backing promising groups and cutting funding to any teams that fail to make significant progress. The focus of the funding would be to make AGI as safe as possible.

3.3.3.3 Differential Technological Progress

Both review boards and government funding could be used to implement "differential intellectual progress"

> Differential intellectual progress consists in prioritizing risk-reducing intellectual progress over risk-increasing intellectual progress. As applied to AI risks in particular, a plan of differential intellectual progress would recommend that our progress on the scientific, philosophical, and technological problems of AI safety outpace our progress on the problems of AI capability such that we develop safe superhuman AIs before we develop (arbitrary) superhuman AIs (Muehlhauser and Salamon 2012).

Examples of research questions that could constitute philosophical or scientific progress in safety can be found in later sections of this paper—for instance, the usefulness of different internal constraints on ensuring safe behavior, or ways of making AGIs reliably adopt human values as they learn what those values are like.

Bostrom (2002) used the term "differential technological progress" to refer to differential intellectual progress in technological development. Bostrom defined differential technological progress as "trying to retard the implementation of dangerous technologies and accelerate implementation of beneficial technologies, especially those that ameliorate the hazards posed by other technologies".

One issue with differential technological progress is that we do not know what kind of progress should be accelerated and what should be retarded. For example, a more advanced communication infrastructure could make AGIs more dangerous, as there would be more networked machines that could be accessed via the Internet. Alternatively, it could be that the world will already be so networked that AGIs will be a major threat anyway, and further advances will make the networks more resilient to attack. Similarly, it can be argued that AGI development is dangerous for as long as we have yet to solve the philosophical problems related to safe AGI design and do not know which AGI architectures are safe to pursue (Yudkowsky 2008). But it can also be argued that we should invest in AGI development now, when the related tools and hardware are still primitive enough that progress will be slow and gradual (Goertzel and Pitt 2012).

3.3.3.4 International Mass Surveillance

For AGI regulation to work, it needs to be enacted on a global scale. This requires solving both the problem of effectively enforcing regulation within a country and the problem of getting many different nations to all agree on the need for regulation.

Shulman (2009) discusses various factors influencing the difficulty of AGI arms control. He notes that AGI technology itself might make international cooperation more feasible. If narrow AIs and early-stage AGIs were used to analyze the information obtained from wide-scale mass surveillance and wiretapping, this might make it easier to ensure that nobody was developing more advanced AGI designs.

Shulman (2010b) similarly notes that machine intelligences could be used to enforce treaties between nations. They could also act as trustworthy inspectors which would be restricted to communicating only information about treaty violations, thus not endangering state secrets even if they were allowed unlimited access to them. This could help establish a "singleton" regulatory regimen capable of effectively enforcing international regulation, including AGI-related treaties. Goertzel and Pitt (2012) also discuss the possibility of having a network of AGIs monitoring the world in order to police other AGIs and to prevent any of them from suddenly obtaining excessive power.

Another proposal for international mass surveillance is to build an "AGI Nanny" (Goertzel 2012b; Goertzel and Pitt 2012), a proposal discussed in Sect. 3.5.4.

Large-scale surveillance efforts are ethically problematic and face major political resistance, and it seems unlikely that current political opinion would support the creation of a far-reaching surveillance network for the sake of AGI risk alone. The extent to which such extremes would be necessary depends on exactly how easy it would be to develop AGI in secret. Although several authors make the point that AGI is much easier to develop unnoticed than something like nuclear weapons (McGinnis 2010; Miller 2012), cutting-edge high-tech research does tend to require major investments which might plausibly be detected even by less elaborate surveillance efforts.

To the extent that surveillance does turn out to be necessary, there is already a strong trend toward a "surveillance society" with increasing amounts of information about people being collected and recorded in various databases (Wood and Ball 2006). As a reaction to the increased surveillance, Mann et al. (2003) propose to counter it with *sous*veillance—giving private individuals the ability to document their life and subject the authorities to surveillance in order to protect civil liberties. This is similar to the proposals of Brin (1998), who argues that technological progress might eventually lead to a "transparent society," where we will need to redesign our societal institutions in a way that allows us to maintain some of our privacy despite omnipresent surveillance. Miller (2012) notes that intelligence agencies are already making major investments in AI-assisted analysis of surveillance data.

If social and technological developments independently create an environment where large-scale surveillance or sousveillance is commonplace, it might be possible to take advantage of those developments in order to police AGI risk.[4] Walker (2008) argues that in order for mass surveillance to become effective, it must be designed in such a way that it will not excessively violate people's privacy, for otherwise the system will face widespread sabotage. Even under such conditions, there is no clear way to define what counts as dangerous AGI. Goertzel and Pitt (2012) point out that there is no clear division between narrow AI and AGI, and attempts to establish such criteria have failed. They argue that since AGI has a nebulous definition, obvious wide-ranging economic benefits, and potentially rich penetration into multiple industry sectors, it is unlikely to be regulated due to speculative long-term risks.

AGI regulation requires global cooperation, as the noncooperation of even a single nation might lead to catastrophe. Historically, achieving global cooperation on tasks such as nuclear disarmament and climate change has been very difficult. As with nuclear weapons, AGI could give an immense economic and military advantage to the country that develops it first, in which case limiting AGI research might even give other countries an incentive to develop AGI faster (Miller 2012).

To be effective, regulation also needs to enjoy support among those being regulated. If developers working in AGI-related fields only follow the letter of the law, while privately viewing all regulations as annoying hindrances, and fears about AGI as overblown, the regulations may prove ineffective. Thus, it might not be

[4]An added benefit would be that this could also help avoid other kinds of existential risk, such as the intentional creation of dangerous new diseases.

enough to convince governments of the need for regulation; the much larger group of people working in the appropriate fields may also need to be convinced.

While Shulman (2009) argues that the unprecedentedly destabilizing effect of AGI could be a cause for world leaders to cooperate more than usual, the opposite argument can be made as well. Gubrud (1997) argues that increased automation could make countries more self-reliant, and international cooperation considerably more difficult. AGI technology is also much harder to detect than, for example, nuclear technology is—nuclear weapons require a substantial infrastructure to develop, while AGI needs much less (McGinnis 2010; Miller 2012).

Miller (2012) even suggests that the mere possibility of a rival being close to developing AGI might, if taken seriously, trigger a nuclear war. The nation that was losing the AGI race might think that being the first to develop AGI was sufficiently valuable that it was worth launching a first strike for, even if it would lose most of its own population in the retaliatory attack. He further argues that, although it would be in the interest of every nation to try to avoid such an outcome, the ease of secretly pursuing an AGI development program undetected, in violation of treaty, could cause most nations to violate the treaty.

Miller also points out that the potential for an AGI arms race exists not only between nations, but between corporations as well. He notes that the more AGI developers there are, the more likely it is that they will all take more risks, with each AGI developer reasoning that if they don't take this risk, somebody else might take that risk first.

Goertzel and Pitt (2012) suggest that for regulation to be enacted, there might need to be an "AGI Sputnik"—a technological achievement that makes the possibility of AGI evident to the public and policy makers. They note that after such a moment, it might not take very long for full human-level AGI to be developed, while the negotiations required to enact new kinds of arms control treaties would take considerably longer.

So far, the discussion has assumed that regulation would be carried out effectively and in the pursuit of humanity's common interests, but actual legislation is strongly affected by lobbying and the desires of interest groups (Olson 1982; Mueller 2003, Chap. 22). Many established interest groups would have an economic interest in either furthering or retarding AGI development, rendering the success of regulation uncertain.

3.3.3.5 "Regulate Research" Proposals—Our View

Although there seem to be great difficulties involved with regulation, there also remains the fact that many technologies have been successfully subjected to international regulation. Even if one were skeptical about the chances of effective regulation, an AGI arms race seems to be one of the worst possible scenarios, one which should be avoided if at all possible. We are therefore generally supportive of regulation, though the most effective regulatory approach remains unclear.

3.3.4 Enhance Human Capabilities

While regulation approaches attempt to limit the kinds of AGIs that will be created, enhancement approaches attempt to give humanity and AGIs a level playing field. In principle, gains in AGI capability would not be a problem if humans could improve themselves to the same level.

Alternatively, human capabilities could be improved in order to obtain a more general capability to deal with difficult problems. Verdoux (2010, 2011) suggests that cognitive enhancement could help in transforming previously incomprehensible mysteries into tractable problems, and Verdoux (2010) particularly highlights the possibility of cognitive enhancement helping to deal with the problems posed by existential risks. One problem with such approaches is that increasing humanity's capability for solving problems will also make it easier to develop dangerous technologies. It is possible that cognitive enhancement should be combined with moral enhancement, in order to help foster the kind of cooperation that would help avoid the risks of technology (Persson and Savulescu 2008, 2012).

Moravec (1988, 1999) proposes that humans could keep up with AGIs via "mind uploading," a process of transferring the information in human brains to computer systems so that human minds could run on a computer substrate. This technology may arrive during a similar timeframe as AGI (Kurzweil 2005; Sandberg and Bostrom 2008; Hayworth 2012; Koene 2012b; Cattell and Parker 2012; Sandberg 2012). However, Moravec argues that mind uploading would come after AGIs, and that unless the uploaded minds ("uploads") would transform themselves to become radically nonhuman, they would be weaker and less competitive than AGIs that were native to a digital environment Moravec (1992, 1999). For these reasons, Warwick (1998) also expresses doubt about the usefulness of mind uploading.

Kurzweil (2005) posits an evolution that will start with brain-computer interfaces, then proceed to using brain-embedded nanobots to enhance our intelligence, and finally lead to full uploading and radical intelligence enhancement. Koene (2012a) criticizes plans to create safe AGIs and considers uploading both a more feasible and a more reliable approach.

Similar proposals have also been made without explicitly mentioning mind uploading. Cade (1966) speculates on the option of gradually merging with machines by replacing body parts with mechanical components. Turney (1991) proposes linking AGIs directly to human brains so that the two meld together into one entity, and Warwick (1998, 2003) notes that cyborgization could be used to enhance humans.

Mind uploading might also be used to make human value systems more accessible and easy to learn for AGIs, such as by having an AGI extrapolate the upload's goals directly from its brain, with the upload providing feedback.

3.3.4.1 Would We Remain Human?

Uploading might destroy parts of humanity that we value (Joy 2000; de Garis 2005). De Garis (2005) argues that a computer could have far more processing power than a human brain, making it pointless to merge computers and humans. The biological component of the resulting hybrid would be insignificant compared to the electronic component, creating a mind that was negligibly different from a "pure" AGI. Kurzweil (2001) makes the same argument, saying that although he supports intelligence enhancement by directly connecting brains and computers, this would only keep pace with AGIs for a couple of additional decades.

The truth of this claim seems to depend on exactly how human brains are augmented. In principle, it seems possible to create a prosthetic extension of a human brain that uses the same basic architecture as the original brain and gradually integrates with it (Sotala and Valpola 2012). A human extending their intelligence using such a method might remain roughly human-like and maintain their original values. However, it could also be possible to connect brains with computer programs that are very unlike human brains, and which would substantially change the way the original brain worked. Even smaller differences could conceivably lead to the adoption of "cyborg values" distinct from ordinary human values (Warwick 2003).

Bostrom (2004) speculates that humans might outsource many of their skills to nonconscious external modules and would cease to experience anything as a result. The value-altering modules would provide substantial advantages to their users, to the point that they could outcompete uploaded minds who did not adopt the modules.

3.3.4.2 Would Evolutionary Pressures Change Us?

A willingness to integrate value-altering modules is not the only way by which a population of uploads might come to have very different values from modern-day humans. This is not necessarily a bad, or even a very novel, development the values of earlier generations have often been different from the values of later generations (Hanson 2012), and it might not be a problem if a civilization of uploads enjoyed very different things than a civilization of humans. Still, as there are possible outcomes that we would consider catastrophic, such as the loss of nearly all things that have intrinsic value for us (Bostrom 2004), it is worth reviewing some of the postulated changes in values.

For comprehensiveness, we will summarize all of the suggested effects that uploading might have on human values, even if they are not obviously negative. Readers may decide for themselves whether or not they consider any of these effects concerning.

Hanson (1994) argues that employers will want to copy uploads who are good workers, and that at least some uploads will consent to being copied in such a manner. He suggests that the resulting evolutionary dynamics would lead to an accelerated evolution of values. This would cause most of the upload population to evolve to be indifferent or favorable to the thought of being copied, to be indifferent toward being deleted as long as another copy of themselves remained, and to be relatively uninterested in having children "the traditional way" (as opposed to copying an already-existing mind). Although Hanson's analysis uses the example of a worker-employer relationship, it should be noted that nations or families, or even single individuals, could also gain a competitive advantage by copying themselves, thus contributing to the strength of the evolutionary dynamic.

Similarly, Bostrom (2004) writes that much of human life's meaning depends on the enjoyment of things ranging from humor and love to literature and parenting. These capabilities were adaptive in our past, but in an upload environment they might cease to be such and gradually disappear entirely.

Shulman (2010b) likewise considers uploading-related evolutionary dynamics. He notes that there might be a strong pressure for uploads to make copies of themselves in such a way that individual copies would be ready to sacrifice themselves to aid the rest. This would favor a willingness to copy oneself, and a view of personal identity which did not consider the loss of a single copy to be death. Beings taking this point of view could then take advantage of economic benefits of continually creating and deleting vast numbers of minds depending on the conditions, favoring the existence of a large number of short-lived copies over a somewhat less efficient world of long-lived minds.

Finally, Sotala and Valpola (2012) consider the possibility of minds coalescing via artificial connections that linked several brains together in the same fashion as the two hemispheres of ordinary brains are linked together. If this were to happen, considerable benefits might accrue to those who were ready to coalesce with other minds. The ability to copy and share memories between minds might also blur distinctions between individual minds. In the end, most humans might cease to be individual, distinct people in any real sense of the word.

It has also been proposed that information security concerns could cause undesirable dynamics among uploads, with significant advantages accruing to those who could steal the computational resources of others and use them to create new copies of themselves. If one could seize control of the hardware that an upload was running on, it could be immediately replaced with a copy of a mind loyal to the attacker. It might even be possible to do this without being detected, if it was possible to steal enough of an upload's personal information to impersonate it.

An attack targeting a critical vulnerability in some commonly used piece of software might quickly hit a very large number of victims. As discussed in the previous chapter, both theoretical arguments and actual cases of malware show that large numbers of machines on the Internet could be infected in a very short time (Staniford et al. 2002, Moore et al. 2002, 2003). In a society of uploads, attacks such as these would be not only inconvenient, but potentially fatal. Eckersley and Sandberg (2013) offer a preliminary analysis of information security in a world of uploads.

3.3.4.3 Would Uploading Help?

Even if the potential changes of values were deemed acceptable, it is unclear whether the technology for uploading could be developed before developing AGI. Uploading might require emulating the low-level details of a human brain with a high degree of precision, requiring large amounts of computing power (Sandberg and Bostrom 2008; Cattell and Parker 2012). In contrast, an AGI might be designed around high-level principles which have been chosen to be computationally cheap to implement on existing hardware architectures.

Yudkowsky (2008) uses the analogy that it is much easier to figure out the principles of aerodynamic flight and then build a Boeing 747 than it is to take a living bird and "upgrade" it into a giant bird that can carry passengers, all while ensuring that the bird remains alive and healthy throughout the process. Likewise, it may be much easier to figure out the basic principles of intelligence and build AGIs than to upload existing minds.

On the other hand, one can also construct an analogy suggesting that it is easier to copy a thing's function than it is to understand how it works. If a person does not understand architecture but wants to build a sturdy house, it may be easier to create a replica of an existing house than it is to design an entirely new one that does not collapse.

Even if uploads were created first, they might not be able to harness all the advantages of digitality, as many of these advantages depend on minds being easy to modify (Sotala 2012), which human minds may not be. Uploads will be able to directly edit their source code as well as introduce simulated pharmaceutical and other interventions, and they could experiment on copies that are restored to an unmodified state if the modifications turn out to be unworkable (Shulman 2010b). Regardless, human brains did not evolve to be easy to modify, and it may be difficult to find a workable set of modifications which would drastically improve them.

In contrast, in order for an AGI programmed using traditional means to be manageable as a software project, it must be easy for the engineers to modify it.[5] Thus, even if uploading were developed before AGI, AGIs that were developed later might still be capable of becoming more powerful than uploads. However, existing uploads already enjoying some of the advantages of the newly-created AGIs would still make it easier for the uploads to control the AGIs, at least for a while.

Moravec (1992) notes that the human mind has evolved to function in an environment which is drastically different from a purely digital environment, and that the only way to remain competitive with AGIs would be to transform into something that was very different from a human. This suggests that uploading

---

[5]However, this might not be true for AGIs created using some alternate means, such as artificial life (Sullins 2005).

might buy time for other approaches, but would be only a short-term solution in and of itself.

If uploading technology were developed before AGI, it could be used to upload a research team or other group and run them at a vastly accelerated rate as compared to the rest of humanity. This would give them a considerable amount of extra time for developing any of the other approaches. If this group were among the first to be successfully emulated and sped up, and if the speed-up would allow enough subjective time to pass before anyone else could implement their own version, they might also be able to avoid trading safety for speed. However, such a group might be able to wield tremendous power, so they would need to be extremely reliable and trustworthy.

3.3.4.4 "Enhance Human Capabilities" Proposals—Our View

Of the various "enhance human capabilities" approaches, uploading proposals seem the most promising, as translating a human brain to a computer program would sidestep many of the constraints that come from modifying a physical system. For example, all relevant brain activity could be recorded for further analysis at an arbitrary level of detail, and any part of the brain could be instantly modified without a need for time-consuming and possibly dangerous invasive surgery. Uploaded brains could also be more easily upgraded to take full advantage of more powerful hardware, while humans whose brains were still partially biological would be bottlenecked by the speed of the biological component.

Uploading does have several problems Uploading research might lead to AGI being created before the uploads, in the long term uploads might have unfavorable evolutionary dynamics, and it seems likely that there will eventually be AGIs which are capable of outperforming uploads in every field of competence. Uploads could also be untrustworthy even without evolutionary dynamics. At the same time, however, uploading research doesn't *necessarily* accelerate AGI research very much, the evolutionary dynamics might not be as bad as they seem at the moment, and the advantages gained from uploading might be enough to help control unsafe AGIs until safe ones could be developed. Methods could also be developed for increasing the trustworthiness of uploads (Shulman 2010b). Uploading might still turn out to be a useful tool for handling AGI risk.

3.3.5 Relinquish Technology

Not everyone believes that the risks involved in creating AGIs are acceptable. *Relinquishment* involves the abandonment of technological development that could lead to AGI. This is possibly the earliest proposed approach, with Butler (1863) writing that "war to the death should be instantly proclaimed" upon machines, for otherwise they would end up destroying humans entirely. In a much-discussed article, Joy (2000) suggests that it might be necessary to relinquish at least some aspects of AGI research, as well as nanotechnology and genetics research.

Hughes (2001) criticizes AGI relinquishment, while Kurzweil (2005) criticizes broad relinquishment but supports the possibility of "fine-grained relinquishment," banning some dangerous aspects of technologies while allowing general work on them to proceed. In general, most writers reject proposals for broad relinquishment.

3.3.5.1 Outlaw AGI

Weng et al. (2009) write that the creation of AGIs would force society to shift from human-centric values to robot-human dual values. In order to avoid this, they consider the possibility of banning AGI. This could be done either permanently or until appropriate solutions are developed for mediating such a conflict of values. McKibben (2003), writing mainly in the context of genetic engineering, also suggests that AGI research should be stopped.

Hughes (2001) argues that attempts to outlaw a technology will only make the technology move to other countries. De Garis (2005) believes that differences of opinion about whether to build AGIs will eventually lead to armed conflict, to the point of open warfare.

Annas et al. (2002) have similarly argued that genetic engineering of humans would eventually lead to war between unmodified humans and the engineered "posthumans," and that cloning and inheritable modifications should therefore be banned. To the extent that one accepts their reasoning with regard to humans, it could also be interpreted to apply to AGIs.

3.3.5.2 Restrict Hardware

Berglas (2012) suggests not only stopping AGI research, but also outlawing the production of more powerful hardware. Berglas holds that it will be possible to build computers as powerful as human brains in the very near future, and that we should therefore reduce the power of new processors and destroy existing ones.[6] Branwen (2012) argues that Moore's Law depends on the existence of a small number of expensive and centralized chip factories, making them easy targets for regulation and incapable of being developed in secret.

[6]Berglas (personal communication) has since changed his mind and no longer believes that it is possible to effectively restrict hardware or otherwise prevent AGI from being created.

3.3.5.3 "Relinquish Technology" Proposals—Our View

Relinquishment proposals suffer from many of the same problems as regulation proposals, but to a greater extent. There is no historical precedent of general, multiuse technology similar to AGI being successfully relinquished for good, nor do there seem to be any theoretical reasons for believing that relinquishment proposals would work in the future. Therefore we do not consider them to be a viable class of proposals.

3.4 External AGI Constraints

Societal approaches involving regulation or research into safe AGI assume that proper AGI design can produce solutions to AGI risk. One category of such solutions is that of *external constraints*. These are restrictions that are imposed on an AGI from the outside and aim to limit its ability to do damage.

Several authors have argued that external constraints are unlikely to work with AGIs that are genuinely far more intelligent than us (Vinge 1993; Yudkowsky 2001; 2008; Kurzweil 2005; Chalmers 2010; Armstrong et al. 2012). The consensus seems to be that external constraints might buy time when dealing with less advanced AGIs, but they are useless against truly superintelligent ones.

External constraints also limit the usefulness of an AGI, as a free-acting one could serve its creators more effectively. This reduces the probability of the universal implementation of external constraints on AGIs. AGIs might also be dangerous if they were confined or otherwise restricted. For further discussion of these points, see Sect. 3.5.1.

3.4.1 AGI Confinement

AGI confinement, or "AI boxing" (Chalmers 2010; Yampolskiy 2012; Armstrong et al. 2012), involves confining an AGI to a specific environment and limiting its access to the external world.

Yampolskiy (2012) makes an attempt to formalize the idea, drawing on previous computer security research on the so-called confinement problem (Lampson 1973). Yampolskiy defines the *AI confinement problem* (AICP) as the challenge of restricting an AGI to a confined environment from which it can't communicate without authorization. A number of methods have been proposed for implementing AI confinement, many of which are extensively discussed in Armstrong et al. (2012).

Chalmers (2010) and Armstrong et al. (2012) mention numerous caveats and difficulties with AI-boxing approaches. A *truly* leakproof system that perfectly isolated the AGI from an outside environment would prevent us from even observing the AGI. If AGIs were given knowledge about human behavior or psychology, they could still launch social engineering attacks on us (Chalmers 2010; Armstrong et al. 2012). An AGI that was unaware of the existence of humans would be less likely to launch such attacks, but also much more limited in the kinds of tasks that it could be used for.

Even if the AGI remained confined, it could achieve enough influence among humans to prevent itself from being reset or otherwise modified (Good 1970). An AGI that people grew reliant on might also become impossible to reset or modify.

3.4.1.1 Safe Questions

Yampolskiy (2012) proposes a hierarchy of security levels for communication with the AGI, ranging from completely unrestricted communication to no communication whatsoever. He suggests that the right level for communicating with AGIs is an intermediate one, based on asking "safe questions." These are defined as questions for which a human could find an answer without help from a superintelligence. For instance, a researcher who was considering two possible cures for cancer could ask the AGI which one to pursue first. Yampolskiy argues that the AGI is effectively not changing the final outcome, as the researcher would in principle be capable of finding the better treatment himself. The AGI, by recommending one of the approaches as the choice to pursue first, is merely speeding up the process. Armstrong et al. (2012) also briefly discuss such approaches.

3.4.1.2 Virtual Worlds

While the confinement protocol supposes an AGI that can perceive our world directly, Chalmers (2010) uses the term "leakproof singularity" to refer to the project of creating isolated virtual worlds which would disallow their inhabitants from influencing the external world. An AGI placed in such a world could be monitored by humans to ensure that it was safe to release into our world. However, Chalmers notes that "for an AI system to be useful or interesting to us at all, it must have some effects on us. At a minimum, we must be able to observe it. And the moment we observe a virtual environment, some information leaks out from that environment into our environment and affects us".

Hibbard (2012b) proposes a decision support system for evaluating AGI designs in simulated environments that model our real environment, and attempts to show that the AGI can be built to be safe.

Armstrong et al. (2012) discuss the notion that we could supplement this approach with several layers of "sub-worlds" virtual worlds stacked within each other. In such a scenario, even if an AGI somehow manages to break out from one virtual reality, it is still restricted by another and cannot directly influence the world. Ultimately, the authors conclude that this approach isn't very promising.

3.4.1.3 Resetting the AGI

For as long as an AGI remains confined or cooperative, its memory could be altered from the outside. Turney (1991) mentions a proposal in which an AGI's memory could be wiped after each job, thus limiting its potential. Armstrong et al. (2012) also mention the possibility of resetting an AGI after each run, though they note that there are many ways for the AGI to get around it, and that it is only useful for certain kinds of AGI systems.

Good (1970) warns that a superhuman AGI may be smart enough to find ways to prevent itself from being successfully reset or shut down, including methods of social manipulation. Warwick (1998) makes a similar point.

3.4.1.4 Checks and Balances

Armstrong et al. (2012) suggest a number of checks and balances that could be used to catch the early stages of an AGI that was behaving dangerously or trying to deceive us. Test questions with known answers could be mixed in with questions we genuinely didn't know the answer to, in order to establish an AGI's trustworthiness. "Honey-pots," tempting resources with great value to the AGI, could also be left in physical or mental locations that it was forbidden from accessing. Finally, several different kinds of AGIs could be presented with the same problem, with their answers then being compared. Inconsistent answers would suggest a possible attempt at deception.

Bostrom (2014, Chap. 9) similarly discusses "tripwires", systems which make diagnostic tests to the AGI and shut it down if they detect signs of dangerous activity. Some possible tripwires includes ones monitoring the AGI's behavior, its abilities, or possibly even content of the AGI's thoughts and planning.

3.4.1.5 "AI Confinement" Proposals—Our View

Despite their limited ability to deal with AGIs more intelligent than us, AI-boxing techniques seem to have value as a first line of defense, and it may be worthwhile to invest in developing off-the-shelf software packages for AI confinement that are easy and cheap to use. A research project that developed AGI unexpectedly might not have been motivated to make major investments in security, but the AGI might still be sufficiently limited in intelligence that confinement would work. Having a defense that is easy to deploy will make it more likely that these kinds of projects will implement better precautions.

However, at the same time there is a risk that this will promote a false sense of security and make research teams think that they have carried out their duty to be cautious merely because they are running elementary confinement protocols. Although some confinement procedures can be implemented on top of an AGI that was not expressly designed for confinement, they are much less reliable than with an AGI that was built with confinement considerations in mind (Armstrong et al. 2012)—and even then, relying solely on confinement is a risky strategy. We are therefore somewhat cautious in our recommendation to develop confinement techniques.

3.4.2 AGI Enforcement

One problem with AI confinement proposals is that humans are tasked with guarding machines that may be far more intelligent than themselves (Good 1970). One proposed solution for this problem is to give the task of watching AGIs to other AGIs.

Armstrong (2007) proposes that the trustworthiness of a superintelligent system could be monitored via a chain of less powerful systems, all the way down to humans. Although humans couldn't verify and understand the workings of a superintelligence, they could verify and understand an AGI just slightly above their own level, which could in turn verify and understand an AGI somewhat above its own level, and so on.

Chaining multiple levels of AI systems with progressively greater capacity seems to be replacing the problem of building a safe AI with a multisystem, and possibly more difficult, version of the same problem. Armstrong himself admits that there are several problems with the proposal. There could be numerous issues along the line, such as a break in the chain of communication or an inability of a system to accurately assess the mind of another (smarter) system. There is also the problem of creating a trusted bottom for the chain in the first place, which is not necessarily any easier than creating a trustworthy superintelligence.

Hall (2007a) writes that there will be a great variety of AGIs, with those that were designed to be moral or aligned with human interests keeping the nonsafe ones in check. Goertzel and Pitt (2012) also propose that we build a community of mutually policing AGI systems of roughly equal levels of intelligence. If an AGI started to "go off the rails," the other AGIs could stop it. This might not prevent a single AGI from undergoing an intelligence explosion, but a community of AGIs might be in a better position to detect and stop it than humans would.

Having AGIs police each other is only useful if the group of AGIs actually has goals and values that are compatible with human goals and values. To this end, the appropriate internal constraints are needed.

The proposal of a society of mutually policing AGIs would avoid the problem of trying to control a more intelligent mind. If a global network of mildly superintelligent AGIs could be instituted in such a manner, it might detect and prevent any nascent takeoff. However, by itself such an approach is not enough to ensure safety

—it helps guard against individual AGIs "going off the rails," but it does not help in a scenario where the programming of *most* AGIs is flawed and leads to nonsafe behavior. It thus needs to be combined with the appropriate internal constraints.

A complication is that a hard takeoff is a *relative* term—an event that happens too fast for any outside observer to stop. Even if the AGI network were a hundred times more intelligent than a network composed of only humans, there might still be a more sophisticated AGI that could overcome the network.

3.4.2.1 "AGI Enforcement" Proposals—Our View

AGI enforcement proposals are in many respects similar to social integration proposals, in that they depend on the AGIs being part of a society which is strong enough to stop any single AGI from misbehaving. The greatest challenge is then to make sure that most of the AGIs in the overall system are safe and do not unite against humans rather than against misbehaving AGIs. Also, there might not be a natural distinction between a distributed AGI and a collection of many different AGIs, and AGI design is in any case likely to make heavy use of earlier AI/AGI techniques. AGI enforcement proposals therefore seem like implementation variants of various internal constraint proposals (Sect. 3.5), rather than independent proposals.

3.5 Internal Constraints

In addition to external constraints, AGIs could be designed with internal motivations designed to ensure that they would take actions in a manner beneficial to humanity. Alternatively, AGIs could be built with internal constraints that make them easier to control via external means.

With regard to internal constraints, Yudkowsky (2008) distinguishes between *technical failure* and *philosophical failure*:

> Technical failure is when you try to build an AI and it doesn't work the way you think it does—you have failed to understand the true workings of your own code. Philosophical failure is trying to build the wrong thing, so that even if you succeeded you would still fail to help anyone or benefit humanity. Needless to say, the two failures are not mutually exclusive (Yudkowsky 2008).

In practice, it is not always easy to distinguish between the two. Most of the discussion below focuses on the philosophical problems of various proposals, but some of the issues, such as whether or not a proposal is actually possible to implement, are technical.

3.5.1 Oracle AI

An *Oracle AI* is a hypothetical AGI that executes no actions other than answering questions. This is a proposal with many similarities to AGI confinement: both involve restricting the extent to which the AGI is allowed to take independent action. We consider the difference to be that an Oracle AI has been programmed to "voluntarily" restrict its activities, whereas AGI confinement refers to methods for restricting an AGI's capabilities even if it was actively attempting to take more extensive action.

Trying to build an AGI that only answered questions might not be as safe as it sounds, however. Correctly defining "take no actions" might prove surprisingly tricky (Armstrong et al. 2012), and the oracle could give flawed advice even if it did correctly restrict its actions.

Some possible examples of flawed advice: As extra resources are useful for the fulfillment of nearly all goals (Omohundro 2007, 2008), the oracle may seek to obtain more resources—such as computing power—in order to answer questions more accurately. Its answers might then be biased toward furthering this goal, even if this temporarily reduces the accuracy of its answers, if it believes this to increase the accuracy of its answers in the long run. Another example is that if the oracle had the goal of answering as many questions as possible as fast as possible, it could attempt to manipulate humans into asking it questions that were maximally simple and easy to answer.

Holden Karnofsky has suggested that an Oracle AI could be safe if it was "just a calculator," a system which only computed things that were asked of it, taking no goal-directed actions of its own. Such a "Tool-Oracle AI" would keep humans as the ultimate decision makers. Furthermore, the first team to create a Tool-Oracle AI could use it to become powerful enough to prevent the creation of other AGIs (Karnofsky and Tallinn 2011; Karnofsky 2012).

An example of a Tool-Oracle AI approach might be Omohundro's (2012) proposal of "Safe-AI Scaffolding" creating highly constrained AGI systems which act within limited, predetermined parameters. These could be used to develop formal verification methods and solve problems related to the design of more intelligent, but still safe, AGI systems.

Oracle AIs might be considered a special case of *domestic AGI* (Bostrom 2014, Chap. 9): AGIs which are built to only be interested in taking action"on a small scale, within a narrow context, and through a limited set of action modes".

3.5.1.1 Oracles Are Likely to Be Released

As with a boxed AGI, there are many factors that would tempt the owners of an Oracle AI to transform it to an autonomously acting agent. Such an AGI would be far more effective in furthering its goals, but also far more dangerous.

Current narrow-AI technology includes high-frequency trading (HFT) algorithms, which make trading decisions within fractions of a second, far too fast to keep humans in the loop. HFT seeks to make a very short-term profit, but even traders looking for a longer-term investment benefit from being faster than their competitors. Market prices are also very effective at incorporating various sources of knowledge (Hanson 2000). As a consequence, a trading algorithm's performance might be improved both by making it faster and by making it more capable of integrating various sources of knowledge. Most advances toward general AGI will likely be quickly taken advantage of in the financial markets, with little opportunity for a human to vet all the decisions. Oracle AIs are unlikely to remain as pure oracles for long.

Similarly, Wallach and Allen (2012) discuss the topic of autonomous robotic weaponry and note that the US military is seeking to eventually transition to a state where the human operators of robot weapons are "on the loop" rather than "in the loop". In other words, whereas a human was previously required to explicitly give the order before a robot was allowed to initiate possibly lethal activity, in the future humans are meant to merely supervise the robot's actions and interfere if something goes wrong. In practice, this may make it much harder for the human to control the robot's actions, if the robot makes a decision which the operator only has a very short time to override.

Docherty & Goose (2012) report on a number of military systems which are becoming increasingly autonomous, with the human oversight for automatic weapons defense systems—designed to detect and shoot down incoming missiles and rockets—already being limited to accepting or overriding the computer's plan of action in a matter of seconds. Although these systems are better described as automatic, carrying out pre-programmed sequences of actions in a structured environment, than autonomous, they are a good demonstration of a situation where rapid decisions are needed and the extent of human oversight is limited. A number of militaries are considering the future use of more autonomous weapons.

In general, any broad domain involving high stakes, adversarial decision making, and a need to act rapidly is likely to become increasingly dominated by autonomous systems. The extent to which the systems will need general intelligence will depend on the domain, but domains such as corporate management, fraud detection, and warfare could plausibly make use of all the intelligence they can get. If one's opponents in the domain are also using increasingly autonomous AI/AGI, there will be an arms race where one might have little choice but to give increasing amounts of control to AI/AGI systems.

Miller (2012) also points out that if a person was close to death, due to natural causes, being on the losing side of a war, or any other reason, they might turn even a potentially dangerous AGI system free. This would be a rational course of action as long as they primarily valued their own survival and thought that even a small chance of the AGI saving their life was better than a near-certain death.

Some AGI designers might also choose to create less constrained and more free-acting AGIs for aesthetic or moral reasons, preferring advanced minds to have more freedom.

3.5.1.2 Oracles Will Become Authorities

Even if humans were technically kept in the loop, they might not have the time, opportunity, motivation, intelligence, or confidence to verify the advice given by an Oracle AI. This may be a danger even with narrower AI systems. Friedman and Kahn (1992) discuss APACHE, an expert system that provides medical advice to doctors. They write that as the medical community puts more and more trust into APACHE, it may become common practice to act automatically on APACHE's recommendations, and it may become increasingly difficult to challenge the "authority" of the recommendations. Eventually, APACHE may in effect begin to dictate clinical decisions.

Likewise, Bostrom and Yudkowsky (2013) point out that modern bureaucrats often follow established procedures to the letter, rather than exercising their own judgment and allowing themselves to be blamed for any mistakes that follow. Dutifully following all the recommendations of an AGI system would be an even better way of avoiding blame.

Wallach and Allen (2012) note the existence of robots which attempt to automatically detect the locations of hostile snipers and to point them out to soldiers. To the extent that these soldiers have come to trust the robots, they could be seen as carrying out the robots' orders. Eventually, equipping the robot with its own weapons would merely dispense with the formality of needing to have a human to pull the trigger.

Thus, even AGI systems that function purely to provide advice will need to be explicitly designed to be safe in the sense of not providing advice that would go against human values (Wallach and Allen 2009). Yudkowsky (2012) further notes that an Oracle AI might choose a plan that is beyond human comprehension, in which case there's still a need to design it as explicitly safe and conforming to human values.

3.5.1.3 "Oracle AI" Proposals—Our View

Much like with external constraints, it seems like Oracle AIs could be a useful stepping stone on the path toward safe, freely acting AGIs. However, because any Oracle AI can be relatively easily turned into a free-acting AGI and because many people will have an incentive to do so, Oracle AIs are not by themselves a solution to AGI risk, even if they are safer than free-acting AGIs when kept as pure oracles.

3.5.2 Top-Down Safe AGI

AGIs built to take autonomous actions will need to be designed with safe motivations. Wallach and Allen divide approaches for ensuring safe behavior into "top-down" and "bottom-up" approaches (Wallach and Allen 2009). They define "top-down" approaches as ones that take a specified ethical theory and attempt to build a system capable of implementing that theory.

They have expressed skepticism about the feasibility of both pure top-down and bottom-up approaches, arguing for a hybrid approach.[7] With regard to top-down approaches, which attempt to derive an internal architecture from a given ethical theory, Wallach (2010) finds three problems

1. "Limitations already recognized by moral philosophers For example, in a utilitarian calculation, how can consequences be calculated when information is limited and the effects of actions cascade in never-ending interactions? Which consequences should be factored into the maximization of utility? Is there a stopping procedure?" (Wallach 2010).
2. The "frame problem" refers to the challenge of discerning relevant from irrelevant information without having to consider all of it, as all information could be relevant in principle (Pylyshyn 1987; Dennett 1987). Moral decision-making involves a number of problems that are related to the frame problem, such as needing to know what effects different actions have on the world, and needing to estimate whether one has sufficient information to accurately predict the consequences of the actions.
3. "The need for background information. What mechanisms will the system require in order to acquire the information it needs to make its calculations? How does one ensure that this information is up to date in real time?" (Wallach 2010).

To some extent, these problems may be special cases of the fact that we do not yet have AGI with good general learning capabilities creating an AGI would also require solving the frame problem, for instance. These problems might therefore not all be as challenging as they seem at first, presuming that we manage to develop AGI in the first place.

3.5.2.1 Three Laws

Probably the most widely known proposal for machine ethics is Isaac Asimov's (1942) Three Laws of Robotics:

1. A robot may not injure a human being or, through inaction, allow a human being to come to harm.
2. A robot must obey orders given to it by human beings except where such orders would conflict with the First Law.
3. A robot must protect its own existence as long as such protection does not conflict with either the First or Second Law.

[7]For a definition of "bottom-up" approaches, see Sect. 5.3.

Asimov and other writers later expanded the list to include a number of additional laws, including the Zeroth Law.

A robot may not harm humanity, or through inaction allow humanity to come to harm.

Although the Three Laws are widely known and have inspired numerous imitations, several of Asimov's own stories were written to illustrate the fact that the laws contained numerous problems. They have also drawn heavy critique from others (Clarke 1993; 1994; Weld and Etzioni 1994; Pynadath and Tambe 2002; Gordon-Spears 2003; McCauley 2007; Weng et al. 2008; Wallach and Allen 2009; Murphy and Woods 2009; Anderson 2011) and are not considered a viable approach for safe AI. Among their chief shortcomings is the fact that they are too ambiguous to implement and, if defined with complete accuracy, contradict each other in many situations.

3.5.2.2 Categorical Imperative

The best-known universal ethical axiom is Kant's categorical imperative. Many authors have discussed using the categorical imperative as the foundation of AGI morality (Stahl 2002; Powers 2006; Wallach and Allen 2009; Beavers 2009, 2012). All of these authors conclude that Kantian ethics is a problematic goal for AGI, though Powers (2006) still remains hopeful about its prospects.

3.5.2.3 Principle of Voluntary Joyous Growth

Goertzel (2004a, b) considers a number of possible axioms before settling on what he calls the "Principle of Voluntary Joyous Growth", defined as "Maximize happiness, growth and choice". He starts by considering the axiom "maximize happiness", but then finds this to be problematic and adds "growth", which he defines as "increase in the amount and complexity of patterns in the universe". Finally he adds "choice" in order to allow sentient beings to "choose their own destiny".

3.5.2.4 Utilitarianism

Classic utilitarianism is an ethical theory stating that people should take actions that lead to the greatest amount of happiness and the smallest amount of suffering. The prospects for AGIs implementing a utilitarian moral theory have been discussed by several authors. The consensus is that pure classical utilitarianism is problematic and does not capture all human values. For example, a purely utilitarian AGI could reprogram the brains of humans so that they did nothing but experience the maximal amount of pleasure all the time, and that prospect seems unsatisfactory to many.[8]

3.5.2.5 Value Learning

Freeman (2009) describes a decision-making algorithm which observes people's behavior, infers their preferences in the form of a utility function, and then attempts to carry out actions which fulfill everyone's preferences. The standard name for this kind of an approach is inverse reinforcement learning (Ng and Russell 2000). Russell (2015) argues for an inverse reinforcement learning approach, as there is considerable data about human behavior and the attitudes behind it, there are solid economic incentives to solve this problem, the problem does not seem intrinsically harder than learning how the rest of the world works, and because it seems possible to define this goal so as to make the AGIs very careful to ensure that they're correct about our preferences before taking any serious action.

Similarly, Dewey (2011) discusses *value learners*, AGIs which are provided a probability distribution over possible utility functions that humans may have. Value learners then attempt to find the utility functions with the best match for human preferences. Hibbard (2012a) builds on Dewey's work to offer a similar proposal.

One problem with conceptualizing human desires as utility functions is that human desires change over time (van Gelder 1995) and also violate the axioms of utility theory required to construct a coherent utility function (Tversky and Kahneman 1981). While it is possible to treat inconsistent choices as random deviations from an underlying "true" utility function that is then learned (Nielsen and Jensen 2004), this does not seem to properly describe human preferences. Rather, human decision making and preferences seem to be driven by multiple competing systems, only some of which resemble utility functions (Dayan 2011). Even if a true utility function could be constructed, it does not take into account the fact that we have second-order preferences, or desires about our desires a drug addict may desire a drug, but also desire that he not desire it (Frankfurt 1971). Similarly, we often wish that we had stronger desires toward behaviors which we consider good but cannot make ourselves engage in. Taking second-order preferences into account leads to what philosophers call "ideal preference" theories of value.

Taking this into account, it has been argued that we should aim to build AGIs which act according to humanity's *extrapolated* values (Yudkowsky 2004; Tarleton 2010; Muehlhauser and Helm 2012). Yudkowsky proposes attempting to discover the "Coherent Extrapolated Volition" (CEV) of humanity, which he defines as

[8]Note that utilitarianism is not the same thing as having a utility function. Utilitarianism is a specific kind of ethical system, while utility functions are general-purpose mechanisms for choosing between actions and can in principle be used to implement very different kinds of ethical systems, such as egoism and possibly even rights-based theories and virtue ethics (Peterson 2010).

> our wish if we knew more, thought faster, were more the people we wished we were, had grown up farther together; where the extrapolation converges rather than diverges, where our wishes cohere rather than interfere; extrapolated as we wish that extrapolated, interpreted as we wish that interpreted (Yudkowsky 2004).

CEV remains vaguely defined and has been criticized by several authors (Hibbard 2005a; Goertzel 2010a; Goertzel and Pitt 2012; Miller 2012). However, Tarleton (2010) finds CEV a promising approach, and suggests that CEV has five desirable properties, and that many different kinds of algorithms could possess these features

- **Meta-algorithm**: Most of the AGI's goals will be obtained at runtime from human minds, rather than explicitly programmed in before runtime.
- **Factually correct beliefs**: The AGI will attempt to obtain correct answers to various factual questions, in order to modify preferences or desires that are based upon false factual beliefs.
- **Singleton**: Only one superintelligent AGI is to be constructed, and it is to take control of the world with whatever goal function is decided upon.
- **Reflection**: Individual or group preferences are reflected upon and revised.
- **Preference aggregation**: The set of preferences of a whole group are to be combined somehow.

At least two CEV variants have been proposed Coherent Aggregated Volition (Goertzel 2010a) and Coherent Blended Volition (Goertzel and Pitt 2012). Goertzel and Pitt (2012) describe a methodology which was used to help end the apartheid in South Africa. The methodology involves people with different views exploring different future scenarios together and in great detail. By exploring different outcomes together, the participants build emotional bonds and mutual understanding, seeking an outcome that everyone can agree to live with. The authors characterize the Coherent Blended Volition of a diverse group as analogous to the "conceptual blend" resulting from the methodology, incorporating the most essential elements of the group into a harmonious whole.

Christiano (2012) attempts to sketch out a formalization of a value extrapolation approach called "indirect normativity". It proposes a technique that would allow an AI to approximate the kinds of values a group of humans would settle on if they could spend an unbounded amount of time and resources considering the problem.

Other authors have begun preliminary work on simpler value learning systems, designed to automatically learn moral principles. Anderson et al. (2005a, 2005b, 2006) have built systems based around various moral duties and principles. As lists of duties do not in and of themselves specify what to do when they conflict, the systems let human experts judge how each conflict should be resolved, and then attempt to learn general rules from the judgments. As put forth, however, this approach would have little ability to infer ethical rules which did not fit the framework of proposed duties. Improved computational models of ethical reasoning, perhaps incorporating work from neuroscience and moral psychology, could help address this. Potapov and Rodionov (2012) propose an approach by which an

AGI could gradually learn the values of other agents as its understanding of the world improved.

A value extrapolation process seems difficult to specify exactly, as it requires building an AGI with programming that formally and rigorously defines human values. Even if it manages to avoid the first issue in Wallach's (2010) list (Sect. 3.5.2), top-down value extrapolation may fall victim to other issues, such as computational tractability. One interpretation of CEV would seem to require modeling not only the values of everyone on Earth, but also the evolution of those values as the people in question interacted with each other, became more intelligent and more like their ideal selves, chose which of their values they wanted to preserve, etc. Even if the AGI could eventually obtain the enormous amount of computing power required to run this model, its behavior would need to be safe from the beginning, or it could end up doing vast damage to humanity before understanding what it was doing wrong.

Goertzel and Pitt's (2012) hybrid approach, in which AGIs cooperate with humans in order to discover the values humans wish to see implemented, seems more likely to avoid the issue of computational tractability. However, it will fail to work in a hard takeoff situation where AGIs take control before being taught the correct human values. Another issue with Coherent Blended Volition is that schemes which require absolute consensus are unworkable with large groups, as anyone whose situation would be worsened by a change of events could block the consensus. A general issue with value extrapolation approaches is that there may be several valid ways of defining a value extrapolation process, with no objective grounds for choosing one rather than another.

Goertzel (2010a) notes that in formal reasoning systems a set of initially inconsistent beliefs which the system attempts to resolve might arrive at something very different than the initial belief set, even if there existed a consistent belief set that was closer to the original set. He suggests that something similar might happen when attempting to make human values consistent, though whether this would be a bad thing is unclear.

3.5.2.6 Approval-Directed Agents

Christiano (2014a, b, c, 2015) proposes "approval-directed agents". Instead of having explicit goals which they would attempt to optimize for, approval-directed agents would choose their next action using a procedure such as

> Estimate the expected rating [an overseer] would give each action if he considered it at length. Take the action with the highest expected rating. (Christiano 2014a, b, c)

Christiano (2014a, b, c) suggests that this approach would:

- allow us to start with simple overseers and simple approval-directed agents, rather than having to specify all of an AI's goals at once

- avoiding getting locked into bad design decisions: "if an approval-directed AI encounters an unforeseen situation, it will respond in the way that we most approve of"
- fail gracefully: flawed interpretations of the overseer's goals might be noticed faster, as the AI wouldn't try to actively hide them unless the overseer deliberately approved of such actions

3.5.2.7 "Top-Down Safe AGI" Proposals—Our View

Of the various top-down proposals, value learning and proposals and approval-directed agents seem to be the only ones which properly take into account the complexity of value thesis (see the pervious chapter), as they attempt to specifically take into account considerations such as "Would humanity have endorsed this course of action if it had known the consequences?" Although there are many open questions concerning the computational tractability as well as the general feasibility of the value learning approaches, they together with the approval-directed ones seem like some of the most important ones to work on.

3.5.3 Bottom-up and Hybrid Safe AGI

Wallach (2010) defines bottom-up approaches as ones that favor evolving or simulating the mechanisms that give rise to our moral decisions. Another alternative is hybrid approaches, combining parts of both top-down and bottom-up approaches.

A problem with pure bottom-up approaches is that techniques such as artificial evolution or merely rewarding the AGI for the right behavior may cause it to behave correctly in tests, but would not guarantee that it would behave safely in any other situation. Even if an AGI *seems* to have adopted human values, the actual processes driving its behavior may be very different from the processes that would be driving the actions of a human who behaved similarly. It might then behave very unexpectedly in situations which are different enough (Yudkowsky 2008, 2011).

Armstrong et al. (2012) discuss various problems related to such approaches and offer examples of concepts which seem straightforward to humans but are not as simple as they may seem on the surface. One of their examples relates to the concept of time

> If the [AGI] had the reasonable-sounding moral premise that "painlessly killing a human being, who is going to die in a micro-second anyway, in order to gain some other good, is not a crime," we would not want it to be able to redefine millennia as seconds (Armstrong et al. 2012).

All humans have an intuitive understanding of time and no experience with beings who could arbitrarily redefine their own clocks and might not share the same concept of time. Such differences in the conceptual grounding of an AGI's values and of human values might not become apparent until too late.

3.5.3.1 Evolutionary Invariants

Human morality is to a large extent shaped by evolution, and evolutionary approaches attempt to replicate this process with AGIs.

Hall (2007a, 2011) argues that self-improving AGIs are likely to exist in competition with many other kinds of self-improving AGIs. Properties that give AGIs a significant disadvantage might then be strongly selected against and disappear. We could attempt to identify *evolutionary invariants*, or evolutionarily stable strategies, which would both survive in a competitive environment and cause an AGI to treat humans well.

Hall (2011) lists self-interest, long planning horizons, knowledge, an understanding of evolutionary ethics, and guaranteed honesty as invariants that are likely to make an AGI more moral as well as to persist even under intense competition. He suggests that, although self-interest may sound like a bad thing in an AGI, non-self-interested creatures are difficult to punish. Thus, enlightened self-interest might be a good thing for an AGI to possess, as it will provide an outside community with both a stick and a carrot to control it with.

Similarly, Waser (2008) suggests that minds which are intelligent enough will, due to game-theoretical and other considerations, become altruistic and cooperative. Waser (2011) proposed the principle of Rational Universal Benevolence (RUB), the idea that the moral course of action is cooperation while letting everyone freely pursue their own goals. Waser proposes that, instead of making human-friendly behavior an AGI's only goal, the AGI would be allowed to have and form its own goals. However, its goals and actions would be subject to the constraint that they should respect the principle of RUB and not force others into a life those others would disagree with.

Kornai (2014) cites Gewirth's (Gewirth 1978) work on the principle of generic consistency, which holds that respecting others' rights to freedom and well-being is a logically necessary conclusion for any rational agents. Kornai suggests that if the principle is correct, then AGIs would respect humanity's rights to freedom and well-being, and that AGIs which failed to respect the principle would be outcompeted by ones which did.

Although these approaches expect AGI either to evolve altruism or to find it the most rational approach, true altruism or even pure tit-for-tat (Axelrod 1987) isn't actually the best strategy in evolutionary terms. Rather, a better strategy is *Machiavellian* tit-for-tat cultivating an appearance of altruism and cooperation when it benefits oneself, and acting selfishly when one can get away with it. Humans seem strongly disposed toward such behavior (Haidt 2006).

Another problem is that tit-for-tat as a good strategy assumes that both players are equally powerful and both have the same options at their disposal. If the AGI became far more powerful than most humans, it might no longer be in its interests to treat humans favorably (Fox and Shulman 2010). This hypothesis can be tested by looking at human behavior if exploiting the weak is an evolutionarily useful strategy, then humans should engage in it when given the opportunity. Humans who feel powerful do indeed devalue the worth of the less powerful and view them as objects of manipulation (Kipnis 1972). They also tend to ignore social norms (van Kleef et al. 2011) and to experience less distress and compassion toward the suffering of others (van-Kleef 2008).

Thus, even if an AGI cooperated with other similarly powerful AGIs, the group of AGIs might still decide to collectively exploit humans. Similarly, even though there might be pressure for AGIs to make themselves more transparent and easily inspected by others, this only persists for as long as the AGI needs others more than the others need the AGI.

3.5.3.2 Evolved Morality

Another proposal is to create AGIs via algorithmic evolution, selecting in each generation the AGIs which are not only the most intelligent, but also the most moral. These ideas are discussed to some extent by Wallach and Allen (2009).

3.5.3.3 Reinforcement Learning

In machine learning, *reinforcement learning* (not to be confused with inverse reinforcement learning, discussed above in the value learning section) is a model in which an agent takes various actions and is differentially rewarded for the actions, after which it learns to perform the actions with the greatest expected reward. In psychology, it refers to agents being rewarded for certain actions and thus learning behaviors which they have a hard time breaking, even if some other kind of behavior is more beneficial for them later on.

Applied to AGI, the machine learning sense of reinforcement involves teaching an AGI to behave in a safe manner by rewarding it for ethical choices, and letting it learn for itself the underlying rules of what constitutes ethical behavior. In an early example of this kind of proposal, McCulloch (1956) described an "ethical machine" that could infer the rules of chess by playing the game, and suggested that it could also learn ethical behavior this way.

Hibbard (2001, 2005a) suggested using reinforcement learning to give AGIs positive emotions toward humans. Early AGIs would be taught to recognize happiness and unhappiness in humans, and the results of this learning would be hard-wired as emotional values in more advanced AGIs. This training process would then be continued—for example, by letting the AGIs predict how human happiness would be affected by various actions and using those predictions as emotional values.

A reinforcement learner is supplied with a reward signal, and it always has the explicit goal of maximizing the sum of this reward, any way it can. In order for this goal to align with human values, humans must engineer the environment so that the reinforcement learner is prevented from receiving rewards if human goals are not fulfilled (Dewey 2011). A reinforcement-learning AGI only remains safe for as long as humans are capable of enforcing this limitation, and will become unpredictable if it becomes capable of overcoming it. Hibbard (Hibbard 2012d) has retracted his earlier reinforcement learning–based proposals, as they would allow the AGI to maximize its reinforcement by modifying humans to be maximally happy, even against their will (Dewey 2011).

3.5.3.4 Human-like AGI

Another kind of proposal involves building AGIs that can learn human values by virtue of being similar to humans.

Connectionist systems, based on artificial neural nets, are capable of learning patterns from data without being told what the patterns are. As some connectionist models have learned to classify problems in a manner similar to humans (McLeod et al. 1998; Plaut 2003; Thomas and McClelland 2008), it has been proposed that connectionist AGI might learn moral principles that are too complex for humans to specify as explicit rules.[9] This idea has been explored by Guarini (2006) and Wallach and Allen (2009).

Sotala (2015) briefly surveys research into human concept learning, and argues that current research suggests that human concepts might be generated using a relatively limited set of mechanisms. He proposes a research program to map these mechanisms in more detail and to then build AGIs which would learn similar concepts as humans, including moral concepts.

One specific proposal that draws upon connectionism is to make AGIs act according to virtue ethics (Wallach and Allen 2009). These authors note that previous writers discussing virtuous behavior have emphasized the importance of learning moral virtues through habit and practice. As it is impossible to exhaustively define a virtue, virtue ethics has traditionally required each individual to learn the right behaviors through "bottom-up processes of discovery or learning". Models that mimicked the human learning process well enough could then potentially learn the same behaviors as humans do.

Another kind of human-inspired proposal is the suggestion that something like Stan Franklin's LIDA architecture (Franklin and Patterson 2006; Ramamurthy et al. 2006; Snaider et al. 2011), or some other approach based on Bernard Baars's (2002, 2005) "global workspace" theory, might enable moral reasoning. In the LIDA architecture, incoming information is monitored by specialized *attention codelets*, each of which searches the input for specific features. In particular, *moral* codelets might look for morally relevant factors and ally themselves with other codelets to

[9]But it should be noted that there are also promising nonconnectionist approaches for modeling human classification behavior—see, e.g., Tenenbaum et al. (2006, 2011).

promote their concerns to the level of conscious attention. Ultimately, some coalitions will win enough support to accomplish a specific kind of decision (Wallach and Allen 2009).

Goertzel and Pitt (2012) consider human memory systems (episodic, sensorimotor, declarative, procedural, attentional, and intentional) and ways by which human morality might be formed via their interaction. They briefly discuss the way that the OpenCog AGI system (Hart and Goertzel 2008; Goertzel 2012a) implements similar memory systems and how those systems could enable it to learn morality. Similarly, Goertzel and Bugaj (2008) discuss the stages of moral development in humans and suggest ways by which they could be replicated in AGI systems, using the specific example of Novamente, a proprietary version of OpenCog.

Waser (2009) also proposes building an AGI by studying the results of evolution and creating an implementation as close to the human model as possible. Shanahan (2015, Chap. 5) similarly suggests building a human-like AGI that might then acquire human values and abilities, such as empathy.

Human-inspired AGI architectures would intuitively seem the most capable of learning human values, though what would be human-like enough remains an open question. It is possible that even a relatively minor variation from the norm could cause an AGI to adopt values that most humans would consider undesirable. Getting the details right might require an extensive understanding of the human brain.

There are also humans who have drastically different ethics than the vast majority of humanity and argue for the desirability of outcomes such as the extinction of mankind (Benatar 2006; Dietrich 2007). There remains the possibility that even AGIs which reasoned about ethics in a completely human-like manner would reach such conclusions.

Humans also have a long track record of abusing power, or of undergoing major behavioral changes due to relatively small injuries—the "safe *Homo sapiens*" problem also remains unsolved. On the other hand, it seems plausible that human-like AGIs could be explicitly engineered to avoid such problems.

The easier that an AGI is to modify, the more powerful it might become (Sotala 2012), and very close recreations of the human brain may turn out to be difficult to extensively modify and upgrade. Human-inspired safe AGIs might then end up outcompeted by AGIs which were easier to modify, and which might or might not be safe.

Even if human-inspired architectures could be easily modified, the messiness of human cognitive architecture means that it might be difficult to ensure that their values remain beneficial during modification. For instance, in LIDA-like architectures, beneficial behavior will depend on the correct coalitions of morality codelets winning each time. If the system undergoes drastic changes, this can be very difficult if not impossible to ensure.

Most AGI builders will attempt to create a mind that displays considerable advantages over ordinary humans. Some such advantages might be best achieved by employing a very nonhuman architecture (Moravec 1992), so there will be reasons to build AGIs that are not as human-like. These could also end up out-competing the human-like AGIs.

3.5.3.5 "Bottom-up and Hybrid Safe AGI" Proposals—Our View

We are generally very skeptical about pure bottom-up methods, as they only allow a very crude degree of control over an AGI's goals, giving it a motivational system which can only be relied on to align with human values in the very specific environments that the AGI has been tested in. Evolutionary invariants seem incapable of preserving complexity of value, and they might not even be capable of preserving human survival. Reinforcement learning, on the other hand, depends on the AGI being incapable of modifying the environment against the will of its human controllers. Therefore, none of these three approaches seems workable.

Human-like AGI might have some promise, depending on exactly how fragile human values were. If the AGI reasoning process could be made sufficiently human-like, there is the possibility that the AGI could remain relatively safe, though less safe than a well-executed value extrapolation–based AGI.

3.5.4 AGI Nanny

A more general proposal, which could be achieved by either top-down, bottom-up, or hybrid methods, is the proposal of an "AGI Nanny" (Goertzel 2012b; Goertzel and Pitt 2012). This is an AGI that is somewhat more intelligent than humans and is designed to monitor Earth for various dangers, including more advanced AGI.

The AGI Nanny would be connected to powerful surveillance systems and would control a massive contingent of robots. It would help abolish problems such as disease, involuntary death, and poverty, while preventing the development of technologies that could threaten humanity. The AGI Nanny would be designed not to rule humanity on a permanent basis, but to give us some breathing room and time to design more advanced AGIs. After some predetermined amount of time, it would cede control of the world to a more intelligent AGI.

Goertzel and Pitt (2012) briefly discuss some of the problems inherent in the AGI Nanny proposal. The AGI would have to come to power in an ethical way, and might behave unpredictably despite our best efforts. It might also be easier to create a dramatically self-improving AGI than to create a more constrained AGI Nanny.

3.5.4.1 "AGI Nanny" Proposals—Our View

Upon asserting control, the AGI Nanny would need to have precisely specified goals, so that it would stop other AGIs from taking control but would also not harm human interests. It is not clear to what extent defining these goals would be easier than defining the goals of a more free-acting AGI (Muehlhauser and Salamon 2012). Overall, the AGI Nanny seems to have promise, but it's unclear whether it can be made to work.

3.5.5 *Motivational Scaffolding*

Motivational scaffolding (Bostrom 2014, Chap. 9) is a proposal in which an AGI is originally given a temporary, relatively simple goal system that is easy to specify but not intended as the final system. As the AGI's capabilities and knowledge mature, the AGI's motivational system is replaced with a more sophisticated one that then guides it on its path towards full superintelligence.

Bostrom (2014) considers this approach to have some potential, but also notes several problems with it. The AGI might resist having its initial goals replaced, the AGI might grow into a superintelligence with the old goals in place, and it is unclear whether installing the final goal system in the fully-developed AGI would be particularly easy.

3.5.6 *Formal Verification*

Formal verification methods prove specific properties about various algorithms. If the complexity and fragility of value theses hold, it follows that safe AGI requires the ability to verify that proposed changes to the AGI will not alter its goals or values. If even a mild drift from an AGI's original goals might lead to catastrophic consequences, then utmost care should be given to ensuring that the goals will not change inadvertently. This is particularly the case if there are no external feedback mechanisms which would correct the drift. Before modifying itself, an AGI could attempt to formally prove that the changes would not alter its existing goals, and would therefore keep them intact even during extended self-modification (Yudkowsky 2008). Such proofs could be required before self-modification was allowed to occur, and the system could also be required to prove that this verify-before-modification property itself would always be preserved during self-modification.

Spears (2006) combines machine learning and formal verification methods to ensure that AIs remain within the bounds of prespecified constraints after having learned new behaviors. She attempts to identify "safe" machine learning operators, which are guaranteed to preserve the system's constraints.

One AGI system built entirely around the concept of formal verification is the Gödel machine (Schmidhuber 2009; Steunebrink and Schmidhuber 2011). It consists of a *solver*, which attempts to achieve the goals set for the machine, and a *searcher*, which has access to a set of axioms that completely describe the machine. The searcher may completely rewrite any part of the machine, provided that it can produce a formal proof showing that such a rewrite will further the system's goals.

Goertzel (2010b) proposes GOLEM (Goal-Oriented LEarning Meta-Architecture), a meta-architecture that can be wrapped around a variety of different AGI systems. GOLEM will only implement changes that are predicted to be more effective at achieving the original goal of the system. Goertzel argues that GOLEM is likely to be both self-improving and steadfast: either it pursues the same goals it had at the start, or it stops acting altogether.

Unfortunately, formalizing the AGI's goals in a manner that will allow formal verification methods to be used is a challenging task. Within cryptography, many communications protocols have been proven secure, only for successful attacks to be later developed against their various implementations. While the formal proofs were correct, they contained assumptions which did not accurately capture the way the protocols worked in practice (Degabriele et al. 2011). Proven theorems are only as good as their assumptions, so formal verification requires good models of the AGI hardware and software.

3.5.6.1 "Formal Verification" Proposals—Our View

Compared to the relatively simple domain of cryptographic security, verifying things such as "Does this kind of a change to the AGI's code preserve its goal of respecting human values?" seems like a much more open-ended and difficult task, one which might even prove impossible. Regardless, it is the only way of achieving high confidence that a system is safe, so it should at least be attempted.

3.5.7 Motivational Weaknesses

Finally, there is a category of internal constraints that, while not making an AGI's *values* safer, make it easier to control AGI via external constraints.

3.5.7.1 High Discount Rates

AGI systems could be given a high discount rate, making them value short-term goals and gains far more than long-term goals and gains (Shulman 2010a; Armstrong et al. 2012). This would inhibit the AGI's long-term planning, making it more predictable. However, an AGI can also reach long-term goals through a series of short-term goals (Armstrong et al. 2012). Another possible problem is that it could cause the AGI to pursue goals which were harmful for humanity's long-term future. Humanity may arguably be seen as already behaving in ways that imply an excessively high discount rate, such as by consuming finite natural resources without properly taking into account the well-being of future generations.

3.5.7.2 Easily Satiable Goals

Shulman (2010a) proposes designing AGIs in such a way that their goals are easy to satisfy. For example, an AGI could receive a near-maximum reward for simply continuing to receive an external reward signal, which could be cut if humans suspected misbehavior. The AGI would then prefer to cooperate with humans rather than trying to attack them, even if it was very sure of its chances of success.[10] Likewise, if the AGI could receive a maximal reward with a relatively small fraction of humanity's available resources, it would have little to gain from seizing more resources.

Bostrom (2014) mentions a variation of this idea, where the AGI would receive a continuous stream of cryptographic reward tokens from a human controller. Its utility function would be defined so that it would receive "99% of its maximum utility from the first reward token; 99% of its remaining utility potential from the second reward token; and so on". Bostrom also discusses what might go wrong with this scheme: possibly the AGI might not trust the human's promises to continue sending the rewards, or it might suspect that the human was in danger of being incapacitated unless the AGI took action.

An extreme form of this kind of a deal is Orseau and Ring's (2011) "Simpleton Gambit," in which an AGI is promised everything that it would ever want, on the condition that it turn itself into a harmless simpleton. Orseau and Ring consider several hypothetical AGI designs, many of which seem likely to accept the gambit, given certain assumptions.

In a related paper, Ring and Orseau (2011) consider the consequences of an AGI being able to modify itself to receive the maximum possible reward. They show that certain types of AGIs will then come to only care about their own survival. Hypothetically, humans could promise not to threaten such AGIs in exchange for them agreeing to be subject to AI-boxing procedures. For this to work, the system would have to believe that humans will take care of its survival against external threats better than it could itself. Hibbard (2012a, c) discusses the kinds of AGIs that would avoid the behaviors described by Ring and Orseau (Ring and Orseau 2011; Orseau and Ring 2011).

3.5.7.3 Calculated Indifference

Another proposal is to make an AGI indifferent to a specific event. For instance, the AGI could be made indifferent to the detonation of explosives attached to its

[10]On the other hand, this might incentivize the AGI to deceive its controllers into believing it was behaving properly, and also to actively hide any information which it even suspected might be interpreted as misbehavior.

hardware, which might enable humans to have better control over it (Armstrong 2010; Armstrong et al. 2012).

3.5.7.4 Programmed Restrictions

Goertzel and Pitt (2012) suggest we ought to ensure that an AGI does not self-improve too fast, because AGIs will be harder to control as they become more and more cognitively superior to humans. To limit the rate of self-improvement in AGIs, perhaps AGIs could be programmed to extensively consult humans and other AGI systems while improving themselves, in order to ensure that no unwanted modifications would be implemented.

Omohundro (2012) discusses a number of programmed restrictions in the form of constraints on what the AGI is allowed to do, with formal proofs being used to ensure that an AGI will not violate its safety constraints. Such limited AGI systems would be used to design more sophisticated AGIs.

Programmed restrictions are problematic, as the AGI might treat these merely as problems to solve in the process of meeting its goals, and attempt to overcome them (Omohundro 2008). Making an AGI not want to quickly self-improve might not solve the problem by itself. If the AGI ends up with a second-order desire to rid itself of such a disinclination, the stronger desire will eventually prevail (Suber 2002). Even if the AGI wanted to maintain its disinclination toward rapid self-improvement, it might still try to circumvent the goal in some other way, such as by creating a copy of itself which did not have that disinclination (Omohundro 2008). Regardless, such limits could help control less sophisticated AGIs.

3.5.7.5 Legal Machine Language

Weng et al. (2008, 2009) propose a "legal machine language" which could be used to formally specify actions which the AGI is allowed or disallowed to do. Governments could then enact laws written in legal machine language, allowing them to be programmed into robots.

3.5.7.6 "Motivational Weaknesses" Proposals—Our View

Overall, motivational weaknesses seem comparable to external constraints possibly useful and worth studying, but not something to rely on exclusively, particularly in the case of superintelligent AGIs. As with external constraints and Oracle AIs, an arms race situation might provide a considerable incentive to loosen or remove such constraints.

3.6 Conclusion

We began this paper by noting that a number of researchers are predicting AGI in the next twenty to one hundred years. One must not put excess trust in this time frame as Armstrong and Sotala (2012) show, experts have been terrible at predicting AGI. Muehlhauser and Salamon (2012) consider a number of methods other than expert opinion that could be used for predicting AGI, but find that they too provide suggestive evidence at best.

It would be a mistake, however, to leap from "AGI is very hard to predict" to "AGI must be very far away". Our brains are known to think about uncertain, abstract ideas like AGI in "far mode," which also makes it feel like AGI must be temporally distant (Trope and Liberman 2010), but something being *uncertain* is not strong evidence that it is *far away*. When we are highly ignorant about something, we should widen our error bars in both directions. Thus, we shouldn't be highly confident that AGI will arrive this century, and we shouldn't be highly confident that it *won't*.

Next, we explained why AGIs may be an existential risk. A trend toward automatization would give AGIs increased influence in society, and there might be a discontinuity in which they gained power rapidly. This could be a disaster for humanity if AGIs don't share our values, and in fact it looks difficult to cause them to share our values because human values are complex and fragile, and therefore problematic to specify.

The recommendations given for dealing with the problem can be divided into proposals for societal action, external constraints, and internal constraints (Table 3.1). Many proposals seem to suffer from serious problems, or seem to be of limited effectiveness. Others seem to have enough promise to be worth exploring. We will conclude by reviewing the proposals which we feel are worthy of further study.

As a brief summary of our views, in the medium term, we think that the proposals of AGI confinement (Sect. 3.4.1), Oracle AI (Sect. 3.5.1), and motivational weaknesses (5.6) would have promise in helping create safer AGIs. These proposals share in common the fact that, although they could help a cautious team of researchers create an AGI, they are not solutions to the problem of AGI risk, as they do not prevent others from creating unsafe AGIs, nor are they sufficient in guaranteeing the safety of sufficiently intelligent AGIs. Regulation (Sect. 3.3.3) as well as human capability enhancement (Sect. 3.3.4) could also help to somewhat reduce AGI risk. In the long run, we will need the ability to guarantee the safety of freely acting AGIs. For this goal, value learning (Sect. 3.5.2.5) would seem like the most reliable approach if it could be made to work, with human-like architecture (Sect. 3.5.3.4) a possible alternative which seems less reliable but possibly easier to build. Formal verification (Sect. 3.5.6) seems like a very important tool in helping to ensure the safety of our AGIs, regardless of the exact approach that we choose.

Of the societal proposals, we are supportive of the calls to regulate AGI development, but we admit there are many practical hurdles which might make this

Table 3.1 Responses to catastrophic AGI risk

Societal proposals			
Do nothing		AGI is distant	
		Little risk, no action needed	
		Let them kill us	
Integrate to society		Legal and economic controls	
		Foster positive values	
Regulate research		Review boards	
		Encourage safety research	
		Differential technological progress	
		International mass surveillance	
Enhance human capabilities			
Relinquish technology		Outlaw AGI	
		Restrict hardware	
AGI design proposals			
External constraints		**Internal constraints**	
AGI confinement	Safe questions	Oracle AI	
	Virtual worlds	Top-down approaches	Three laws
	Resetting the AGI		Categorical imperative
	Checks and balances		Principle of Voluntary Joyous Growth
AGI enforcement			Utilitarianism
			Value learning
			Approval-directed agents
		Bottom-up and hybrid approaches	Evolutionary invariants
			Evolved morality
			Reinforcement learning
			Human-like AGI
		AGI Nanny	
		Formal verification	
		Motivational scaffolding	
		Motivational weaknesses	High discount rates
			Easily satiable goals
			Calculated indifference
			Programmed restrictions
			Legal Machine Language

infeasible. The economic and military potential of AGI, and the difficulty of enforcing compliance to regulations and arms treaties could lead to unstoppable arms races.

We find ourselves in general agreement with the authors who advocate funding additional research into safe AGI as the primary solution. Such research will also help establish the kinds of constraints which would make it possible to successfully carry out integration proposals.

Uploading approaches, in which human minds are made to run on computers and then augmented, might buy us some time to develop safe AGI. However, it is unclear whether they can be developed before AGI, and large-scale uploading could create strong evolutionary trends which seem dangerous in and of themselves. As AGIs seem likely to eventually outpace uploads, uploading by itself is probably not a sufficient solution. What uploading could do is to reduce the initial advantages that AGIs enjoy over (partially uploaded) humanity, so that other responses to AGI risk can be deployed more effectively.

External constraints are likely to be useful in controlling AGI systems of limited intelligence, and could possibly help us develop more intelligent AGIs while maintaining their safety. If inexpensive external constraints were readily available, this could encourage even research teams skeptical about safety issues to implement them. Yet it does not seem safe to rely on these constraints once we are dealing with a superhuman intelligence, and we cannot trust everyone to be responsible enough to contain their AGI systems, especially given the economic pressures to "release" AGIs. For such an approach to be a solution for AGI risk in general, it would have to be adopted by all successful AGI projects, at least until safe AGIs were developed. Much the same is true of attempting to design Oracle AIs. In the short term, such efforts may be reinforced by research into motivational weaknesses, internal constraints that make AGIs easier to control via external means.

In the long term, the internal constraints that show the most promise are value extrapolation approaches and human-like architectures. Value extrapolation attempts to learn human values and interpret them as we would wish them to be interpreted. These approaches have the advantage of potentially maximizing the preservation of human values, and the disadvantage that such approaches may prove intractable, or impossible to properly formalize. Human-like architectures seem easier to construct, as we can simply copy mechanisms that are used within the human brain, but it seems hard to build such an exact match as to reliably replicate human values. Slavish reproductions of the human psyche also seem likely to be outcompeted by less human, more efficient architectures.

Both approaches would benefit from better formal verification methods, so that AGIs which were editing and improving themselves could verify that the modifications did not threaten to remove the AGIs' motivation to follow their original goals. Studies which aim to uncover the roots of human morals and preferences also seem like candidates for research that would benefit the development of safe AGI, as do studies into computational models of ethical reasoning.

We reiterate that when we talk about "human values", we are not making the claim that human values would be static, nor that *current* human values would be ideal. Nor do we wish to imply that the values of other sentient beings would be unimportant. Rather, we are seeking to guarantee the implementation of some very basic values, such as the avoidance of unnecessary suffering, the preservation of humanity, and the prohibition of forced brain reprogramming. We believe the vast majority of humans would agree with these values and be sad to see them lost.

Acknowledgementss Special thanks to Luke Muehlhauser for extensive assistance throughout the writing process. We would also like to thank Abram Demski, Alexei Turchin, Alexey Potapov, Anders Sandberg, Andras Kornai, Anthony Berglas, Aron Vallinder, Ben Goertzel, Ben Noble, Ben Sterrett, Brian Rabkin, Bill Hibbard, Carl Shulman, Dana Scott, Daniel Dewey, David Pearce, Evelyn Mitchell, Evgenij Thorstensen, Frank White, gwern branwen, Harri Valpola, Jaan Tallinn, Jacob Steinhardt, James Babcock, James Miller, Joshua Fox, Louie Helm, Mark Gubrud, Mark Waser, Michael Anissimov, Michael Vassar, Miles Brundage, Moshe Looks, Randal Koene, Robin Hanson, Risto Saarelma, Steve Omohundro, Suzanne Lidström, Steven Kaas, Stuart Armstrong, Tim Freeman, Ted Goertzel, Toni Heinonen, Tony Barrett, Vincent Müller, Vladimir Nesov, Wei Dai, and two anonymous reviewers as well as several users of lesswrong.com for their helpful comments.

References

Agliata, Daniel, and Stacey Tantleff-Dunn. 2004. "The Impact of Media Exposure on Males' Body Image". Journal of Social and Clinical Psychology 23(1): 7–22. doi:10.1521/jscp.23.1.7.26988.

Alexander, Scott. 2015. "AI researchers on AI risk". Slate Star Codex [blog]. http://slatestarcodex.com/2015/05/22/ai-researchers-on-ai-risk/.

Anderson, Monica. 2010. "Problem Solved: Unfriendly AI". H + Magazine, December 15. http://hplusmagazine.com/2010/12/15/problem-solved-unfriendly-ai/.

Anderson, Michael, Susan Leigh Anderson, and Chris Armen, eds. 2005a. Machine Ethics: Papers from the 2005 AAAI Fall Symposium. Technical Report, FS-05-06. AAAI Press, Menlo Park, CA. http://www.aaai.org/Library/Symposia/Fall/fs05-06.

Anderson, Michael, Susan Leigh Anderson, and Chris Armen. 2005b. "MedEthEx: Toward a Medical Ethics Advisor." In Caring Machines: AI in Eldercare: Papers from the 2005 AAAI Fall Symposium, edited by Timothy Bickmore, 9–16. Technical Report, FS-05-02. AAAI Press, Menlo Park, CA. http://aaaipress.org/Papers/Symposia/Fall/2005/FS-05-02/FS05-02-002.pdf.

Anderson, Michael, Susan Leigh Anderson, and Chris Armen. 2006. "An Approach to Computing Ethics." IEEE Intelligent Systems 21(4): 56–63. doi: 10.1109/MIS.2006.64.

Anderson, Susan Leigh. 2011. "The Unacceptability of Asimov's Three Laws of Robotics as a Basis for Machine Ethics". In Anderson and Anderson 2011, 285–296.

Annas, George J., Lori B. Andrews, and Rosario M. Isasi. 2002. "Protecting the Endangered Human: Toward an International Treaty Prohibiting Cloning and Inheritable Alterations". American Journal of Law & Medicine 28(2–3): 151–178.

Anthony, Dick, and Thomas Robbins. 2004. "Conversion and 'Brainwashing' in New Religious Movements". In The Oxford Handbook of New Religious Movements, 1st ed., edited by James R. Lewis, 243–297. New York: Oxford University Press. doi:10.1093/oxfordhb/9780195369649.003. 0012.

Armstrong, Stuart. 2007. "Chaining God: A Qualitative Approach to AI, Trust and Moral Systems". Unpublished manuscript, October 20. Accessed December 31, 2012. http://www.neweuropeancentury.org/GodAI.pdf.

Armstrong, Stuart. 2010. Utility Indifference. Technical Report, 2010-1. Oxford: Future of Humanity Institute, University of Oxford. http://www.fhi.ox.ac.uk/reports/2010-1.pdf.

Armstrong, Stuart, Anders Sandberg, and Nick Bostrom. 2012. "Thinking Inside the Box: Controlling and Using an Oracle AI". Minds and Machines 22(4): 299–324. doi:10.1007/s11023-012-9282-2.

Armstrong, Stuart, and Kaj Sotala. 2012. "How We're Predicting AI — or Failing To". In Beyond AI: Artificial Dreams, edited by Jan Romportl, Pavel Ircing, Eva Zackova, Michal Polak, and Radek Schuster, 52–75. Pilsen: University of West Bohemia. Accessed February 2, 2013. http://www.kky.zcu.cz/en/publications/1/JanRomportl_2012_BeyondAIArtificial.pdf.

Asimov, Isaac. 1942. "Runaround". Astounding Science-Fiction, March, 94–103.
Axelrod, Robert. 1987. "The Evolution of Strategies in the Iterated Prisoner's Dilemma". In Genetic Algorithms and Simulated Annealing, edited by Lawrence Davis, 32–41. Los Altos, CA: Morgan Kaufmann.
Baars, Bernard J. 2002. "The Conscious Access Hypothesis: Origins and Recent Evidence". Trends in Cognitive Sciences 6(1): 47–52. doi:10.1016/S1364-6613(00)01819-2.
Baars, Bernard J. 2005. "Global Workspace Theory of Consciousness: Toward a Cognitive Neuroscience of Human Experience". In The Boundaries of Consciousness: Neurobiology and Neuropathology, edited by Steven Laureys, 45–53. Progress in Brain Research 150. Boston: Elsevier.
Beavers, Anthony F. 2009. "Between Angels and Animals: The Question of Robot Ethics; or, Is Kantian Moral Agency Desirable?" Paper presented at the Annual Meeting of the Association for Practical and Professional Ethics, Cincinnati, OH, March.
Beavers, Anthony F. 2012. "Moral Machines and the Threat of Ethical Nihilism". In Lin, Patrick, Keith Abney, and George A. Bekey, eds. Robot Ethics: The Ethical and Social Implications of Robotics. Intelligent Robotics and Autonomous Agents. Cambridge, MA: MIT Press, 333–344.
Benatar, David. 2006. Better Never to Have Been: The Harm of Coming into Existence. New York: Oxford University Press.
Berglas, Anthony. 2012. "Artificial Intelligence Will Kill Our Grandchildren (Singularity)". Unpublished manuscript, draft 9, January. Accessed December 31, 2012. http://berglas.org/Articles/AIKillGrandchildren/AIKillGrandchildren.html.
Bostrom, Nick. 2002. "Existential Risks: Analyzing Human Extinction Scenarios and Related Hazards." Journal of Evolution and Technology 9. http://www.jetpress.org/volume9/risks.html.
Bostrom, Nick. 2004. "The Future of Human Evolution". In Two Hundred Years After Kant, Fifty Years After Turing, edited by Charles Tandy, 339–371. Vol. 2. Death and Anti-Death. Palo Alto, CA: Ria University Press.
Bostrom, Nick. 2012. "The Superintelligent Will: Motivation and Instrumental Rationality in Advanced Artificial Agents". In "Theory and Philosophy of AI," edited by Vincent C. Müller. Special issue, Minds and Machines 22(2): 71–85. doi:10.1007/s11023-012-9281-3.
Bostrom, Nick. 2014. Superintelligence: Paths, dangers, strategies. Oxford University Press.
Bostrom, Nick, and Eliezer Yudkowsky. 2013. "The Ethics of Artificial Intelligence". In Cambridge Handbook of Artificial Intelligence, edited by Keith Frankish and William Ramsey. New York: Cambridge University Press.
Branwen, Gwern. 2012. "Slowing Moore's Law: Why You Might Want to and How You Would Do It". gwern.net. December 11. Accessed December 31, 2012. http://www.gwern.net/Slowing%20Moore's%20Law.
Brin, David. 1998. The Transparent Society: Will Technology Force Us to Choose Between Privacy and Freedom? Reading, MA: Perseus Books.
Bringsjord, Selmer, and Alexander Bringsjord. 2012. "Belief in the Singularity is Fideistic". In Eden, Amnon, Johnny Søraker, James H. Moor, and Eric Steinhart, eds. Singularity Hypotheses: A Scientific and Philosophical Assessment. The Frontiers Collection. Berlin: Springer.
Brooks, Rodney A. 2008. "I, Rodney Brooks, Am a Robot". IEEE Spectrum 45(6): 68–71. doi:10.1109/MSPEC.2008.4531466.
Brynjolfsson, Erik, and Andrew McAfee. 2011. Race Against The Machine: How the Digital Revolution is Accelerating Innovation, Driving Productivity, and Irreversibly Transforming Employment and the Economy. Lexington, MA: Digital Frontier. Kindle edition.
Bryson, Joanna, and Phil Kime. 1998. "Just Another Artifact: Ethics and the Empirical Experience of AI". Paper presented at the Fifteenth Internation Congress on Cybernetics, Namur, Belgium. http://www.cs.bath.ac.uk/~jjb/web/aiethics98.html.
Butler, Samuel [Cellarius, pseud.]. 1863. "Darwin Among the Machines". Christchurch Press, June 13. http://www.nzetc.org/tm/scholarly/tei-ButFir-t1-g1-t1-g1-t4-body.html.
Cade, C. Maxwell. 1966. Other Worlds Than Ours. 1st ed. London: Museum.

Cattell, Rick, and Alice Parker. 2012. Challenges for Brain Emulation: Why is Building a Brain so Difficult? Synaptic Link, February 5. http://synapticlink.org/Brain%20Emulation%20Challenges.pdf.

Chalmers, David John. 2010. "The Singularity: A Philosophical Analysis". Journal of Consciousness Studies 17 (9–10): 7–65. http://www.ingentaconnect.com/content/imp/jcs/2010/00000017/f0020009/art00001.

Christiano, Paul F. 2012. "'Indirect Normativity' Write-up". Ordinary Ideas (blog), April 21. http://ordinaryideas.wordpress.com/2012/04/21/indirect-normativity-write-up/.

Christiano, Paul F. 2014a Approval-directed agents. December 1 https://medium.com/ai-control/model-free-decisions-6e6609f5d99e.

Christiano, Paul F. 2014b. Approval-directed search. December 14 https://medium.com/@paulfchristiano/approval-directed-search-63457096f9e4.

Christiano, Paul F. 2014c. Approval-directed bootstrapping. December 20 https://medium.com/ai-control/approval-directed-bootstrapping-5d49e886c14f.

Christiano, Paul F. 2015. Learn policies or goals? April 21 https://medium.com/ai-control/learn-policies-or-goals-348add76b8eb.

Clark, Gregory. 2007. A Farewell to Alms: A Brief Economic History of the World. 1st ed. Princeton, NJ: Princeton University Press.

Clarke, Roger. 1993. "Asimov's Laws of Robotics: Implications for Information Technology, Part 1". Computer 26(12): 53–61. doi:10.1109/2.247652.

Clarke, Roger. 1994. "Asimov's Laws of Robotics: Implications for Information Technology, Part 2". Computer 27 (1): 57–66. doi:10.1109/2.248881.

Daley, William. 2011. "Mitigating Potential Hazards to Humans from the Development of Intelligent Machines". Synthese 2:44–50. http://www.synesisjournal.com/vol2_g/2011_2_44-50_Daley.pdf.

Davis, Ernest. 2012. "The Singularity and the State of the Art in Artificial Intelligence". Working Paper, New York, May 9. Accessed July 22, 2013. http://www.cs.nyu.edu/~davise/papers/singularity.pdf.

Dayan, Peter. 2011. "Models of Value and Choice". In Neuroscience of Preference and Choice: Cognitive and Neural Mechanisms, edited by Raymond J. Dolan and Tali Sharot, 33–52. Waltham, MA: Academic Press.

De Garis, Hugo. 2005. The Artilect War: Cosmists vs. Terrans: A Bitter Controversy Concerning Whether Humanity Should Build Godlike Massively Intelligent Machines.Palm Springs, CA: ETC Publications.

Degabriele, Jean Paul, Kenny Paterson, and Gaven Watson. 2011. "Provable Security in the Real World". IEEE Security & Privacy Magazine 9(3): 33–41. doi:10.1109/MSP.2010.200.

Dennett, Daniel C. 1987. "Cognitive Wheels: The Frame Problem of AI". In Pylyshyn 1987, 41–64.

Dennett, Daniel C. 2012. "The Mystery of David Chalmers". Journal of Consciousness Studies 19 (1–2): 86–95. http://ingentaconnect.com/content/imp/jcs/2012/00000019/F0020001/art00005.

Deutsch, David. 2011. The Beginning of Infinity: Explanations that Transform the World. 1st ed. New York: Viking.

Dewey, Daniel. 2011. "Learning What to Value". In Schmidhuber, Jürgen, Kristinn R. Thórisson, and Moshe Looks, eds. Artificial General Intelligence: 4th International Conference, AGI 2011, Mountain View, CA, USA, August 3–6, 2011. Proceedings. Lecture Notes in Computer Science 6830. Berlin: Springer, 309–314.

Dietrich, Eric. 2007. "After The Humans Are Gone". Philosophy Now, May–June. http://philosophynow.org/issues/61/After_The_Humans_Are_Gone.

Docherty, Bonnie, and Steve Goose. 2012. Losing Humanity: The Case Against Killer Robots. Cambridge, MA: Human Rights Watch and the International Human Rights Clinic, November 19. http://www.hrw.org/sites/default/files/reports/arms1112ForUpload_0_0.pdf.

Douglas, Thomas. 2008. "Moral Enhancement". Journal of Applied Philosophy 25(3): 228–245. doi:10.1111/j.1468-5930.2008.00412.x.

Eckersley, Peter, and Anders Sandberg. 2013. Is Brain Emulation Dangerous? Journal of Artificial General Intelligence 4.3: 170–194.
Fox, Joshua, and Carl Shulman. 2010. "Superintelligence Does Not Imply Benevolence". In Mainzer, Klaus, ed. 2010. ECAP10: VIII European Conference on Computing and Philosophy. Munich: Dr. Hut.
Frankfurt, Harry G. 1971. "Freedom of the Will and the Concept of a Person". Journal of Philosophy 68 (1): 5–20. doi:10.2307/2024717.
Franklin, Stan, and F. G. Patterson Jr. 2006. "The LIDA Architecture: Adding New Modes of Learning to an Intelligent, Autonomous, Software Agent". In IDPT-2006 Proceedings.San Diego, CA: Society for Design & Process Science. http://ccrg.cs.memphis.edu/assets/papers/zo-1010-lida-060403.pdf.
Freeman, Tim. 2009. "Using Compassion and Respect to Motivate an Artificial Intelligence". Unpublished manuscript, March 8. Accessed December 31, 2012. http://fungible.com/respect/paper.html.
Friedman, Batya, and Peter H. Kahn. 1992. "Human Agency and Responsible Computing: Implications for Computer System Design". Journal of Systems and Software 17 (1): 7–14. doi:10.1016/0164-1212(92)90075-U.
Future of Life Institute. 2015. Research Priorities for Robust and Beneficial Artificial Intelligence: an Open Letter. http://futureoflife.org/misc/open_letter.
Gewirth, Alan. 1978. Reason and Morality. Chicago: University of Chicago Press.
Goertzel, Ben. 2004a. "Encouraging a Positive Transcension: Issues in Transhumanist Ethical Philosophy". Dynamical Psychology.http://www.goertzel.org/dynapsyc/2004/PositiveTranscension.htm.
Goertzel, Ben. 2004b. "Growth, Choice and Joy: Toward a Precise Definition of a Universal Ethical Principle". Dynamical Psychology. http://www.goertzel.org/dynapsyc/2004/GrowthChoiceJoy.htm.
Goertzel, Ben. 2010a. "Coherent Aggregated Volition: A Method for Deriving Goal System Content for Advanced, Beneficial AGIs". The Multiverse According to Ben (blog), March 12. http://multiverseaccordingtoben.blogspot.ca/2010/03/coherent-aggregated-volitiontoward.html.
Goertzel, Ben. 2010b. "GOLEM: Toward an AGI Meta-Architecture Enabling Both Goal Preservation and Radical Self-Improvement". Unpublished manuscript, May 2. Accessed December 31, 2012. http://goertzel.org/GOLEM.pdf.
Goertzel, Ben. 2012a. "CogPrime: An Integrative Architecture for Embodied Artificial General Intelligence". OpenCog Foundation. October 2. Accessed December 31, 2012. http://wiki.opencog.org/w/CogPrime_Overview.
Goertzel, Ben. 2012b. "Should Humanity Build a Global AI Nanny to Delay the Singularity Until It's Better Understood?" Journal of Consciousness Studies 19(1–2): 96–111. http://ingentaconnect.com/content/imp/jcs/2012/00000019/F0020001/art00006.
Goertzel, Ben, and Stephan Vladimir Bugaj. 2008. "Stages of Ethical Development in Artificial General Intelligence Systems". In Wang, Pei, Ben Goertzel, and Stan Franklin, eds. Artificial General Intelligence 2008: Proceedings of the First AGI Conference. Frontiers in Artificial Intelligence and Applications 171. Amsterdam: IOS, 448–459.
Goertzel, Ben, and Joel Pitt. 2012. "Nine Ways to Bias Open-Source AGI Toward Friendliness". Journal of Evolution and Technology 22(1): 116–131. http://jetpress.org/v22/goertzel-pitt.htm.
Gomes, Lee. 2015. Facebook AI Director Yann LeCun on His Quest to Unleash Deep Learning and Make Machines Smarter. IEEE Spectrum. http://spectrum.ieee.org/automaton/robotics/artificial-intelligence/facebook-ai-director-yann-lecun-on-deep-learning#qaTopicEight.
Good, Irving John. 1970. "Some Future Social Repercussions of Computers". International Journal of Environmental Studies 1(1–4): 67–79. doi:10.1080/00207237008709398.
Gordon-Spears, Diana F. 2003. "Asimov's Laws: Current Progress". In Formal Approaches to Agent-Based Systems: Second International Workshop, FAABS 2002, Greenbelt, MD, USA, October 29–31, 2002. Revised Papers, edited by Michael G. Hinchey, James L. Rash, Walter F. Truszkowski, Christopher Rouff, and Diana F. Gordon-Spears, 257–259. Lecture Notes in Computer Science 2699. Berlin: Springer. doi:10.1007/978-3-540-45133-4_23.

Groesz, Lisa M., Michael P. Levine, and Sarah K. Murnen. 2001. "The Effect of Experimental Presentation of Thin Media Images on Body Satisfaction: A Meta-Analytic Review". International Journal of Eating Disorders 31(1): 1–16. doi:10.1002/eat.10005.

Guarini, Marcello. 2006. "Particularism and the Classification and Reclassification of Moral Cases". IEEE Intelligent Systems 21 (4): 22–28. doi:10.1109/MIS.2006.76.

Gubrud, Mark Avrum. 1997. "Nanotechnology and International Security". Paper presented at the Fifth Foresight Conference on Molecular Nanotechnology, Palo Alto, CA, November 5–8. http://www.foresight.org/Conferences/MNT05/Papers/Gubrud/.

Gunkel, David J. 2012. The Machine Question: Critical Perspectives on AI, Robotics, and Ethics. Cambridge, MA: MIT Press.

Haidt, Jonathan. 2006. The Happiness Hypothesis: Finding Modern Truth in Ancient Wisdom. 1st ed. New York: Basic Books.

Hall, John Storrs. 2007a. Beyond AI: Creating the Conscience of the Machine. Amherst, NY: Prometheus Books.

Hall, John Storrs. 2011. "Ethics for Self-Improving Machines". In Anderson and Anderson 2011, 512– 523.

Hanson, Robin. 1994. "If Uploads Come First: The Crack of a Future Dawn". Extropy 6(2). http://hanson.gmu.edu/uploads.html.

Hanson, Robin. 2000. "Shall We Vote on Values, But Bet on Beliefs?" Unpublished manuscript, September. Last revised October 2007. http://hanson.gmu.edu/futarchy.pdf.

Hanson, Robin. 2008. "Economics of the Singularity". IEEE Spectrum 45 (6): 45–50. doi:10.1109/MSPEC.2008.4531461.

Hanson, Robin. 2009. "Prefer Law to Values". Overcoming Bias (blog), October 10. http://www.overcomingbias.com/2009/10/prefer-law-to-values.html.

Hanson, Robin. 2012. "Meet the New Conflict, Same as the Old Conflict". Journal of Consciousness Studies 19(1–2): 119–125. http://www.ingentaconnect.com/content/imp/jcs/2012/00000019/F0020001/art00008.

Hare, Robert D., Danny Clark, Martin Grann, and David Thornton. 2000. "Psychopathy and the Predictive Validity of the PCL-R: An International Perspective". Behavioral Sciences & the Law 18(5): 623–645. doi:10.1002/1099-0798(200010)18:5<623::AID-BSL409>3.0.CO;2-W.

Harris, Grant T., and Marnie E. Rice. 2006. "Treatment of Psychopathy: A Review of Empirical Findings". In Handbook of Psychopathy, edited by Christopher J. Patrick, 555–572. New York: Guilford.

Hart, David, and Ben Goertzel. 2008. "OpenCog: A Software Framework for Integrative Artificial General Intelligence". Unpublished manuscript. http://www.agiri.org/OpenCog_AGI-08.pdf.

Hayworth, Kenneth J. 2012. "Electron Imaging Technology for Whole Brain Neural Circuit Mapping". International Journal of Machine Consciousness 4(1): 87–108. doi:10.1142/S1793843012500060.

Heylighen, Francis. 2007. "Accelerating Socio-Technological Evolution: From Ephemeralization and Stigmergy to the Global Brain". In Globalization as Evolutionary Process: Modeling Global Change, edited by George Modelski, Tessaleno Devezas, and William R. Thompson, 284–309. Rethinking Globalizations 10. New York: Routledge.

Heylighen, Francis. 2012. "Brain in a Vat Cannot Break Out." Journal of Consciousness Studies 19 (1–2): 126–142. http://www.ingentaconnect.com/content/imp/jcs/2012/00000019/F0020001/art00009.

Hibbard, Bill. 2001. "Super-Intelligent Machines". ACM SIGGRAPH Computer Graphics 35 (1): 13–15. http://www.siggraph.org/publications/newsletter/issues/v35/v35n1.pdf.

Hibbard, Bill. 2005a. "Critique of the SIAI Collective Volition Theory". Unpublished manuscript, December. Accessed December 31, 2012. http://www.ssec.wisc.edu/~billh/g/SIAI_CV_critique.html.

Hibbard, Bill. 2005b. "The Ethics and Politics of Super-Intelligent Machines". Unpublished manuscript, July. Microsoft Word file, accessed December 31, 2012. https://sites.google.com/site/whibbard/g/SI_ethics_politics.doc.

Hibbard, Bill. 2008. "Open Source AI." In Wang, Pei, Ben Goertzel, and Stan Franklin, eds. Artificial General Intelligence 2008: Proceedings of the First AGI Conference. Frontiers in Artificial Intelligence and Applications 171. Amsterdam: IOS, 473–477.
Hibbard, Bill. 2012a. "Avoiding Unintended AI Behaviors". In Bach, Joscha, Ben Goertzel, and Matthew Iklé, eds. Artificial General Intelligence: 5th International Conference, AGI 2012, Oxford, UK, December 8–11, 2012. Proceedings. Lecture Notes in Artificial Intelligence 7716. New York: Springer. doi: 10.1007/978-3-642-35506-6, 107–116.
Hibbard, Bill. 2012b. "Decision Support for Safe AI Design". In Bach, Joscha, Ben Goertzel, and Matthew Iklé, eds. Artificial General Intelligence: 5th International Conference, AGI 2012, Oxford, UK, December 8–11, 2012. Proceedings. Lecture Notes in Artificial Intelligence 7716. New York: Springer. doi: 10.1007/978-3-642-35506-6, 117–125.
Hibbard, Bill. 2012c. "Model-Based Utility Functions". Journal of Artificial General Intelligence 3 (1): 1–24. doi:10.2478/v10229-011-0013-5.
Hibbard, Bill. 2012d. The Error in My 2001 VisFiles Column, September. Accessed December 31, 2012. http://www.ssec.wisc.edu/~billh/g/visfiles_error.html.
Horvitz, Eric J., and Bart Selman. 2009. Interim Report from the AAAI Presidential Panel on Long- Term AI Futures. Palo Alto, CA: AAAI, August. http://www.aaai.org/Organization/Panel/panelnote.pdf.
Hughes, James. 2001. "Relinquishment or Regulation: Dealing with Apocalyptic Technological Threats". Hartford, CT, November 14.
IEEE Spectrum. 2008. "Tech Luminaries Address Singularity": "The Singularity; Special Report". (June).
Jenkins, Anne. 2003. "Artificial Intelligence and the Real World". Futures 35 (7): 779–786. doi:10.1016/S0016-3287(03)00029-6.
Joy, Bill. 2000. "Why the Future Doesn't Need Us". Wired, April. http://www.wired.com/wired/archive/8.04/joy.html.
Karnofsky, Holden. 2012. "Thoughts on the Singularity Institute (SI)". Less Wrong (blog), May 11. http://lesswrong.com/lw/cbs/thoughts_on_the_singularity_institute_si/.
Karnofsky, Holden, and Jaan Tallinn. 2011. "Karnofsky & Tallinn Dialog on SIAI Efficacy". Accessed December 31, 2012. http://xa.yimg.com/kq/groups/23070378/1331435883/name/Jaan+Tallinn+2011+05+-+revised.doc.
Kipnis, David. 1972. "Does Power Corrupt?". Journal of Personality and Social Psychology 24(1): 33–41. doi:10.1037/h0033390.
Koene, Randal A. 2012a. "Embracing Competitive Balance: The Case for Substrate-Independent Minds and Whole Brain Emulation". In Eden, Amnon, Johnny Søraker, James H. Moor, and Eric Steinhart, eds. Singularity Hypotheses: A Scientific and Philosophical Assessment. The Frontiers Collection. Berlin: Springer.
Koene, Randal A. 2012b. "Experimental Research in Whole Brain Emulation: The Need for Innovative in Vivo Measurement Techniques". International Journal of Machine Consciousness 4(1): 35–65. doi:10.1142/S1793843012400033.
Kornai, András. 2014. Bounding the impact of AGI. Journal of Experimental & Theoretical Artificial Intelligence, 26(3), 417–438.
Kurzweil, Ray. 2001. "Response to Stephen Hawking". Kurzweil Accelerating Intelligence. September 5. Accessed December 31, 2012. http://www.kurzweilai.net/response-to-stephen-hawking.
Kurzweil, Ray. 2005. The Singularity Is Near: When Humans Transcend Biology. New York: Viking.
Lampson, Butler W. 1973. "A Note on the Confinement Problem". Communications of the ACM 16(10): 613–615. doi:10.1145/362375.362389.
Legg, Shane. 2009. "Funding Safe AGI". Vetta Project (blog), August 3. http://www.vetta.org/2009/08/funding-safe-agi/.
Madrigal, Alexis C. 2015. The case against killer robots, from a guy actually working on artificial intelligence. http://fusion.net/story/54583/the-case-against-killer-robots-from-a-guy-actually-building-ai/.

Mann, Steve, Jason Nolan, and Barry Wellman. 2003. "Sousveillance: Inventing and Using Wearable Computing Devices for Data Collection in Surveillance Environments". Surveillance & Society 1(3): 331–355. http://library.queensu.ca/ojs/index.php/surveillance-and-society/article/view/3344.

McCauley, Lee. 2007. "AI Armageddon and the Three Laws of Robotics". Ethics and Information Technology 9(2): 153–164. doi:10.1007/s10676-007-9138-2.

McCulloch, W. S. 1956. "Toward Some Circuitry of Ethical Robots; or, An Observational Science of the Genesis of Social Evaluation in the Mind-like Behavior of Artifacts". Acta Biotheoretica 11(3–4): 147–156. doi:10.1007/BF01557008.

McDermott, Drew. 2012. "Response to 'The Singularity' by David Chalmers". Journal of Consciousness Studies 19(1–2): 167–172. http://www.ingentaconnect.com/content/imp/jcs/2012/00000019/F0020001/art00011.

McGinnis, John O. 2010. "Accelerating AI". Northwestern University Law Review 104 (3): 1253–1270. http://www.law.northwestern.edu/lawreview/v104/n3/1253/LR104n3McGinnis.pdf.

McKibben, Bill. 2003. Enough: Staying Human in an Engineered Age. New York: Henry Holt.

McLeod, Peter, Kim Plunkett, and Edmund T. Rolls. 1998. Introduction to Connectionist Modelling of Cognitive Processes. New York: Oxford University Press.

Miller, James D. 2012. Singularity Rising: Surviving and Thriving in a Smarter, Richer, and More Dangerous World. Dallas, TX: BenBella Books.

Moore, David, Vern Paxson, Stefan Savage, Colleen Shannon, Stuart Staniford, and Nicholas Weaver. 2003. "Inside the Slammer Worm". IEEE Security & Privacy Magazine 1(4): 33–39. doi:10.1109/MSECP.2003.1219056.

Moore, David, Colleen Shannon, and Jeffery Brown. 2002. "Code-Red: A Case Study on the Spread and Victims of an Internet Worm". In Proceedings of the Second ACM SIGCOMM Workshop on Internet Measurment (IMW'02), 273–284. New York: ACM Press. doi:10.1145/637201.637244.

Moravec, Hans P. 1988. Mind Children: The Future of Robot and Human Intelligence. Cambridge, MA: Harvard University Press.

Moravec, Hans P. 1992. "Pigs in Cyberspace". Field Robotics Center. Accessed December 31, 2012. http://www.frc.ri.cmu.edu/~hpm/project.archive/general.articles/1992/CyberPigs.html.

Moravec, Hans P. 1999. Robot: Mere Machine to Transcendent Mind. New York: Oxford University Press.

Muehlhauser, Luke, and Louie Helm. 2012. "The Singularity and Machine Ethics". In Eden, Amnon, Johnny Søraker, James H. Moor, and Eric Steinhart, eds. Singularity Hypotheses: A Scientific and Philosophical Assessment. The Frontiers Collection. Berlin: Springer.

Muehlhauser, Luke, and Anna Salamon. 2012. "Intelligence Explosion: Evidence and Import". In Eden, Amnon, Johnny Søraker, James H. Moor, and Eric Steinhart, eds. Singularity Hypotheses: A Scientific and Philosophical Assessment.The Frontiers Collection. Berlin: Springer.

Mueller, Dennis C. 2003. Public Choice III. 3rd ed. New York: Cambridge University Press.

Müller, Vincent C., and Nick Bostrom. 2014. Future progress in artificial intelligence: A survey of expert opinion. Fundamental Issues of Artificial Intelligence.

Murphy, Robin, and David D. Woods. 2009. "Beyond Asimov: The Three Laws of Responsible Robotics". IEEE Intelligent Systems 24(4): 14–20. doi:10.1109/MIS.2009.69.

Napier, William. 2008. "Hazards from Comets and Asteroids". In Bostrom, Nick, and Milan M. Ćirković, eds. Global Catastrophic Risks. New York: Oxford University Press, 222–237.

Ng, Andrew Y., and Stuart J. Russell. 2000. Algorithms for inverse reinforcement learning. In Icml (pp. 663–670).

Nielsen, Thomas D., and Finn V. Jensen. 2004. "Learning a Decision Maker's Utility Function from (Possibly) Inconsistent Behavior". Artificial Intelligence 160(1–2): 53–78. doi:10.1016/j.artint.2004.08.003.

Nordmann, Alfred. 2007. "If and Then: A Critique of Speculative NanoEthics". NanoEthics 1(1): 31–46. doi:10.1007/s11569-007-0007-6.

Nordmann, Alfred. 2008. "Singular Simplicity". IEEE Spectrum, June. http://spectrum.ieee.org/robotics/robotics-software/singular-simplicity.
Olson, Mancur. 1982. The Rise and Decline of Nations: Economic Growth, Stagflation, and Social Rigidities. New Haven, CT: Yale University Press.
Omohundro, Stephen M. 2007. "The Nature of Self-Improving Artificial Intelligence". Paper presented at Singularity Summit 2007, San Francisco, CA, September 8–9. http://selfaware systems.com/2007/10/05/paper-on-the-nature-of-self-improving-artificial-intelligence/.
Omohundro, Stephen M. 2008. "The Basic AI Drives". In Wang, Pei, Ben Goertzel, and Stan Franklin, eds. Artificial General Intelligence 2008: Proceedings of the First AGI Conference. Frontiers in Artificial Intelligence and Applications 171. Amsterdam: IOS, 483–492.
Omohundro, Stephen M. 2012. "Rational Artificial Intelligence for the Greater Good". In Eden, Amnon, Johnny Søraker, James H. Moor, and Eric Steinhart, eds. Singularity Hypotheses: A Scientific and Philosophical Assessment. The Frontiers Collection. Berlin: Springer.
Orseau, Laurent, and Mark Ring. 2011. "Self-Modification and Mortality in Artificial Agents". In Schmidhuber, Jürgen, Kristinn R. Thórisson, and Moshe Looks, eds. Artificial General Intelligence: 4th International Conference, AGI 2011, Mountain View, CA, USA, August 3–6, 2011. Proceedings. Lecture Notes in Computer Science 6830. Berlin: Springer, 1–10.
Persson, Ingmar, and Julian Savulescu. 2008. "The Perils of Cognitive Enhancement and the Urgent Imperative to Enhance the Moral Character of Humanity". Journal of Applied Philosophy 25(3): 162– 177. doi:10.1111/j.1468-5930.2008.00410.x.
Persson, Ingmar, and Julian Savulescu. 2012. Unfit for the Future. Oxford: Oxford University Press. doi:10.1093/acprof:oso/9780199653645.001.0001.
Peterson, Nathaniel R., David B. Pisoni, and Richard T. Miyamoto. 2010. "Cochlear Implants and Spoken Language Processing Abilities: Review and Assessment of the Literature". Restorative Neurology and Neuroscience 28(2): 237–250. doi:10.3233/RNN-2010-0535.
Plaut, David C. 2003. "Connectionist Modeling of Language: Examples and Implications". In Mind, Brain, and Language: Multidisciplinary Perspectives, edited by Marie T. Banich and Molly Mack, 143–168. Mahwah, NJ: Lawrence Erlbaum.
Posner, Richard A. 2004. Catastrophe: Risk and Response. New York: Oxford University Press.
Potapov, Alexey, and Sergey Rodionov. 2012. "Universal Empathy and Ethical Bias for Artificial General Intelligence". Paper presented at the Fifth Conference on Artificial General Intelligence (AGI– 12), Oxford, December 8–11. Accessed June 27, 2013. http://aideus.com/research/doc/preprints/04_paper4_AGIImpacts12.pdf.
Powers, Thomas M. 2006. "Prospects for a Kantian Machine". IEEE Intelligent Systems 21(4): 46–51. doi:10.1109/MIS.2006.77.
Pylyshyn, Zenon W., ed. 1987. The Robot's Dilemma: The Frame Problem in Artificial Intelligence. Norwood, NJ: Ablex.
Pynadath, David V., and Milind Tambe. 2002. "Revisiting Asimov's First Law: A Response to the Call to Arms". In Intelligent Agents VIII: Agent Theories, Architectures, and Languages 8th International Workshop, ATAL 2001 Seattle, WA, USA, August 1–3, 2001 Revised Papers, edited by John-Jules Ch. Meyer and Milind Tambe, 307–320. Berlin: Springer. doi:10.1007/3-540-45448-9_22.
Ramamurthy, Uma, Bernard J. Baars, Sidney K. D'Mello, and Stan Franklin. 2006. "LIDA: A Working Model of Cognition". In Proceedings of the Seventh International Conference on Cognitive Modeling, edited by Danilo Fum, Fabio Del Missier, and Andrea Stocco, 244–249. Trieste, Italy: Edizioni Goliardiche. http://ccrg.cs.memphis.edu/assets/papers/ICCM06-UR.pdf.
Ring, Mark, and Laurent Orseau. 2011. "Delusion, Survival, and Intelligent Agents". In Schmidhuber, Jürgen, Kristinn R. Thórisson, and Moshe Looks, eds. Artificial General Intelligence: 4th International Conference, AGI 2011, Mountain View, CA, USA, August 3–6, 2011. Proceedings. Lecture Notes in Computer Science 6830. Berlin: Springer, 11–20.
Russell, Stuart J. 2015. Will They Make Us Better People? Edge.org. http://edge.org/response-detail/26157.

Russell, Stuart J., Dewey, Daniel, Tegmark, Max. 2015. Research priorities for robust and beneficial artificial intelligence. http://futureoflife.org/static/data/documents/research_priorities.pdf.

Sandberg, Anders. 2001. "Friendly Superintelligence". Accessed December 31, 2012. http://www.aleph.se/Nada/Extro5/Friendly%20Superintelligence.htm.

Sandberg, Anders. 2012. "Models of a Singularity". In Eden, Amnon, Johnny Søraker, James H. Moor, and Eric Steinhart, eds. Singularity Hypotheses: A Scientific and Philosophical Assessment. The Frontiers Collection. Berlin: Springer.

Sandberg, Anders, and Nick Bostrom. 2008. Whole Brain Emulation: A Roadmap. Technical Report, 2008-3. Future of Humanity Institute, University of Oxford. http://www.fhi.ox.ac.uk/wpcontent/uploads/brain-emulation-roadmap-report1.pdf.

Schmidhuber, Jürgen. 2009. "Ultimate Cognition à la Gödel". Cognitive Computation 1(2): 177–193. doi:10.1007/s12559-009-9014-y.

Scott, James C. 1998. Seeing Like a State: How Certain Schemes to Improve the Human Condition Have Failed. New Haven, CT: Yale University Press.

Shanahan, Murray. 2015. The Technological Singularity. MIT Press (forthcoming).

Shulman, Carl. 2009. "Arms Control and Intelligence Explosions". Paper presented at the 7th European Conference on Computing and Philosophy (ECAP), Bellaterra, Spain, July 2–4.

Shulman, Carl. 2010a. Omohundro's "Basic AI Drives" and Catastrophic Risks. The Singularity Institute, San Francisco, CA. http://intelligence.org/files/BasicAIDrives.pdf.

Shulman, Carl. 2010b. Whole Brain Emulation and the Evolution of Superorganisms. The Singularity Institute, San Francisco, CA. http://intelligence.org/files/WBE-Superorgs.pdf.

Snaider, Javier, Ryan Mccall, and Stan Franklin. 2011. "The LIDA Framework as a General Tool for AGI". In Schmidhuber, Jürgen, Kristinn R. Thórisson, and Moshe Looks, eds. Artificial General Intelligence: 4th International Conference, AGI 2011, Mountain View, CA, USA, August 3–6, 2011. Proceedings. Lecture Notes in Computer Science 6830. Berlin: Springer, 133–142.

Soares, N., & Benja Fallenstein. 2014. Aligning Superintelligence with Human Interests: A Technical Research Agenda. Tech. rep. Machine Intelligence Research Institute, 2014. URL: http://intelligence.org/files/TechnicalAgenda.pdf.

Sobolewski, Matthias. 2012. "German Cabinet to Agree Tougher Rules on High-Frequency Trading". Reuters, September 25. Accessed December 31, 2012. http://in.reuters.com/article/2012/09/25/germany-bourse-rules-idINL5E8KP8BK20120925.

Sotala, Kaj. 2012. "Advantages of Artificial Intelligences, Uploads, and Digital Minds". International Journal of Machine Consciousness 4(1): 275–291. doi:10.1142/S1793843012400161.

Sotala, Kaj. 2015. Concept learning for safe autonomous AI. In Workshops at the Twenty-Ninth AAAI Conference on Artificial Intelligence.

Sotala, Kaj, and Harri Valpola. 2012. "Coalescing Minds: Brain Uploading-Related Group Mind Scenarios". International Journal of Machine Consciousness 4(1): 293–312. doi:10.1142/S1793843012400173.

Sotala, Kaj, and Roman V. Yampolskiy. 2013. Responses to catastrophic AGI risk: a survey. Technical report 2013-2. Berkeley, CA: Machine Intelligence Research Institute.

Sotala, Kaj, and Roman V. Yampolskiy. 2015. Responses to catastrophic AGI risk: a survey. Physica Scripta, 90(1), 018001.

Spears, Diana F. 2006. "Assuring the Behavior of Adaptive Agents". In Agent Technology from a Formal Perspective, edited by Christopher Rouff, Michael Hinchey, James Rash, Walter Truszkowski, and Diana F. Gordon-Spears, 227–257. NASA Monographs in Systems and Software Engineering. London: Springer. doi:10.1007/1-84628-271-3_8.

Stahl, Bernd Carsten. 2002. "Can a Computer Adhere to the Categorical Imperative? A Contemplation of the Limits of Transcendental Ethics in IT". In, edited by Iva Smit and George E. Lasker, 13–18. Vol. 1. Windsor, ON: International Institute for Advanced Studies in Systems Research/Cybernetics.

Staniford, Stuart, Vern Paxson, and Nicholas Weaver. 2002. "How to 0wn the Internet in Your Spare Time". In Proceedings of the 11th USENIX Security Symposium, edited by Dan Boneh, 149–167. Berkeley, CA: USENIX. http://www.icir.org/vern/papers/cdc-usenix-sec02/.

Steunebrink, Bas R., and Jürgen Schmidhuber. 2011. "A Family of Gödel Machine Implementations". In Schmidhuber, Jürgen, Kristinn R. Thórisson, and Moshe Looks, eds. Artificial General Intelligence: 4th International Conference, AGI 2011, Mountain View, CA, USA, August 3–6, 2011. Proceedings. Lecture Notes in Computer Science 6830. Berlin: Springer, 275–280.

Suber, Peter. 2002. "Saving Machines from Themselves: The Ethics of Deep Self-Modification". Accessed December 31, 2012. http://www.earlham.edu/~peters/writing/selfmod.htm.

Sullins, John P. 2005. "Ethics and Artificial life: From Modeling to Moral Agents". Ethics & Information Technology 7 (3): 139–148. doi:10.1007/s10676-006-0003-5.

Tarleton, Nick. 2010. Coherent Extrapolated Volition: A Meta-Level Approach to Machine Ethics. The Singularity Institute, San Francisco, CA. http://intelligence.org/files/CEV-MachineEthics.pdf.

Tenenbaum, Joshua B., Thomas L. Griffiths, and Charles Kemp. 2006. "Theory-Based Bayesian Models of Inductive Learning and Reasoning". In "Probabilistic Models of Cognition". Special issue, Trends in Cognitive Sciences 10(7): 309–318. doi:10.1016/j.tics.2006.05.009.

Tenenbaum, Joshua B., Charles Kemp, Thomas L. Griffiths, and Noah D. Goodman. 2011. "How to grow a mind: Statistics, structure, and abstraction". science 331, 6022: 1279–1285.

Thomas, Michael S. C., and James L. McClelland. 2008. "Connectionist Models of Cognition". In The Cambridge Handbook of Computational Psychology, edited by Ron Sun, 23–58. Cambridge Handbooks in Psychology. New York: Cambridge University Press.

Trope, Yaacov, and Nira Liberman. 2010. "Construal-level Theory of Psychological Distance". Psychological Review 117(2): 440–463. doi:10.1037/a0018963.

Turney, Peter. 1991. "Controlling Super-Intelligent Machines". Canadian Artificial Intelligence, July 27, 3–4, 12, 35.

Tversky, Amos, and Daniel Kahneman. 1981. "The Framing of Decisions and the Psychology of Choice". Science 211 (4481): 453–458. doi:10.1126/science.7455683.

Van Gelder, Timothy. 1995. "What Might Cognition Be, If Not Computation?" Journal of Philosophy 92(7): 345–381. http://www.jstor.org/stable/2941061.

Van Kleef, Gerben A., Astrid C. Homan, Catrin Finkenauer, Seval Gundemir, and Eftychia Stamkou. 2011. "Breaking the Rules to Rise to Power: How Norm Violators Gain Power in the Eyes of Others". Social Psychological and Personality Science 2(5): 500–507. doi:10.1177/1948550611398416.

Van Kleef, Gerben A., Christopher Oveis, Ilmo van der Löwe, Aleksandr LuoKogan, Jennifer Goetz, and Dacher Keltner. 2008. "Power, Distress, and Compassion: Turning a Blind Eye to the Suffering of Others". Psychological Science 19(12): 1315–1322. doi:10.1111/j.1467-9280.2008.02241.x.

Verdoux, Philippe. 2010. "Risk Mysterianism and Cognitive Boosters". Journal of Futures Studies 15 (1): 1–20. Accessed February 2, 2013. http://www.jfs.tku.edu.tw/15-1/A01.pdf.

Verdoux, Philippe. 2011. "Emerging Technologies and the Future of Philosophy". Metaphilosophy 42(5): 682–707. doi:10.1111/j.1467-9973.2011.01715.x.

Vinge, Vernor. 1993. "The Coming Technological Singularity: How to Survive in the Post-Human Era". In Vision-21: Interdisciplinary Science and Engineering in the Era of Cyberspace, 11–22. NASA Conference Publication 10129. NASA Lewis Research Center. http://ntrs.nasa.gov/archive/nasa/casi.ntrs.nasa.gov/19940022855_1994022855.pdf.

Walker, Mark. 2008. "Human Extinction and Farsighted Universal Surveillance". Working Paper, September. Accessed December 31, 2012. http://www.nmsu.edu/~philos/documents/sept-2008-smart-dust-final.doc.

Wallach, Wendell. 2010. "Robot Minds and Human Ethics: The Need for a Comprehensive Model of Moral Decision Making". In "Robot Ethics and Human Ethics," edited by Anthony Beavers. Special issue, Ethics and Information Technology 12(3): 243–250. doi:10.1007/s10676-010-9232-8.

Wallach, Wendell, and Colin Allen. 2009. Moral Machines: Teaching Robots Right from Wrong. New York: Oxford University Press. doi:10.1093/acprof:oso/9780195374049.001.0001.
Wallach, Wendell, and Colin Allen. 2012. "Framing Robot Arms Control". Ethics and Information Technology. doi:10.1007/ s10676-012-9303-0.
Wang, Pei. 2012. "Motivation Management in AGI Systems". In Bach, Joscha, Ben Goertzel, and Matthew Iklé, eds. Artificial General Intelligence: 5th International Conference, AGI 2012, Oxford, UK, December 8–11, 2012. Proceedings. Lecture Notes in Artificial Intelligence 7716. New York: Springer. doi: 10.1007/978-3-642-35506-6, 352–361.
Warwick, Kevin. 1998. In the Mind of the Machine: Breakthrough in Artificial Intelligence. London: Arrow.
Warwick, Kevin. 2003. "Cyborg Morals, Cyborg Values, Cyborg Ethics". Ethics and Information Technology 5(3): 131–137. doi:10.1023/B:ETIN.0000006870.65865.cf.
Waser, Mark R. 2008. "Discovering the Foundations of a Universal System of Ethics as a Road to Safe Artificial Intelligence". In Biologically Inspired Cognitive Architectures: Papers from the AAAI Fall Symposium, 195–200. Technical Report, FS-08-04. AAAI Press, Menlo Park, CA. http://www.aaai.org/Papers/Symposia/Fall/2008/FS-08-04/FS08-04-049.pdf.
Waser, Mark R. 2009. "A Safe Ethical System for Intelligent Machines". In Biologically Inspired Cognitive Architectures: Papers from the AAAI Fall Symposium, edited by Alexei V. Samsonovich, 194–199. Technical Report, FS-09-01. AAAI Press, Menlo Park, CA. http://aaai.org/ocs/index.php/FSS/FSS09/paper/view/934.
Waser, Mark R. 2011. "Rational Universal Benevolence: Simpler, Safer, and Wiser than 'Friendly AI'". In Schmidhuber, Jürgen, Kristinn R. Thórisson, and Moshe Looks, eds. Artificial General Intelligence: 4th International Conference, AGI 2011, Mountain View, CA, USA, August 3–6, 2011. Proceedings. Lecture Notes in Computer Science 6830. Berlin: Springer, 153–162.
Weld, Daniel, and Oren Etzioni. 1994. "The First Law of Robotics (A Call to Arms)". In Proceedings of the Twelfth National Conference on Artificial Intelligence, edited by Barbara Hayes-Roth and Richard E. Korf, 1042–1047. Menlo Park, CA: AAAI Press. http://www.aaai.org/Papers/AAAI/1994/AAAI94-160.pdf.
Weng, Yueh-Hsuan, Chien-Hsun Chen, and Chuen-Tsai Sun. 2008. "Safety Intelligence and Legal Machine Language: Do We Need the Three Laws of Robotics?" In Service Robot Applications, edited by Yoshihiko Takahashi. InTech. doi:10.5772/6057.
Weng, Yueh-Hsuan, Chien-Hsun Chen, and Chuen-Tsai Sun. 2009. "Toward the Human–Robot Coexistence Society: On Safety Intelligence for Next Generation Robots". International Journal of Social Robotics 1(4): 267–282. doi:10.1007/s12369-009-0019-1.
Whitby, Blay. 1996. Reflections on Artificial Intelligence: The Legal, Moral, and Ethical Dimensions. Exeter, UK: Intellect Books.
Whitby, Blay, and Kane Oliver. 2000. "How to Avoid a Robot Takeover: Political and Ethical Choices in the Design and Introduction of Intelligent Artifacts". Paper presented at Symposium on Artificial Intelligence, Ethics and (Quasi-) Human Rights at AISB-00, University of Birmingham, England. http://www.sussex.ac.uk/Users/blayw/BlayAISB00.html.
Wilson, Grant. 2013. Minimizing global catastrophic and existential risks from emerging technologies through international law. Va. Envtl. LJ, 31, 307.
Wood, David Murakami, and Kirstie Ball, eds. 2006. A Report on the Surveillance Society: For the Information Commissioner, by the Surveillance Studies Network. Wilmslow, UK: Office of the Information Commissioner, September. http://www.ico.org.uk/about_us/research/~/media/documents/library/Data_Protection/Practical_application/SURVEILLANCE_SOCIETY_SUMMARY_06.ashx.
Yampolskiy, Roman V. 2012. "Leakproofing the Singularity: Artificial Intelligence Confinement Problem". Journal of Consciousness Studies 2012(1–2): 194–214. http://www.ingentaconnect.com/content/imp/jcs/2012/00000019/F0020001/art00014.
Yampolskiy, Roman V. 2013. "What to Do with the Singularity Paradox?" Studies in Applied Philosophy, Epistemology and Rational Ethics vol 5, pp. 397–413. Springer Berlin Heidelberg.
Yampolskiy, Roman V., and Joshua Fox. 2012. "Safety Engineering for Artificial General Intelligence". Topoi. doi:10.1007/s11245-012-9128-9.

Yudkowsky, Eliezer. 2001. Creating Friendly AI 1.0: The Analysis and Design of Benevolent Goal Architectures. The Singularity Institute, San Francisco, CA, June 15. http://intelligence.org/files/CFAI.pdf.
Yudkowsky, Eliezer. 2004. Coherent Extrapolated Volition. The Singularity Institute, San Francisco, CA, May. http://intelligence.org/files/CEV.pdf.
Yudkowsky, Eliezer. 2008. "Artificial Intelligence as a Positive and Negative Factor in Global Risk". In Bostrom, Nick, and Milan M. Ćirković, eds. Global Catastrophic Risks. New York: Oxford University Press, 308–345.
Yudkowsky, Eliezer. 2011. Complex Value Systems are Required to Realize Valuable Futures. The Singularity Institute, San Francisco, CA. http://intelligence.org/files/ComplexValues.pdf.
Yudkowsky, Eliezer. 2012. "Reply to Holden on 'Tool AI'". Less Wrong (blog), June 12. http://lesswrong.com/lw/cze/reply_to_holden_on_tool_ai/.

Part II
Managing the Singularity Journey

Chapter 4
How Change Agencies Can Affect Our Path Towards a Singularity

Ping Zheng and Mohammed-Asif Akhmad

4.1 Introduction

This chapter adopts a change agency perspective to help us understand how technological change may mutually interact with human agency and socioeconomic forces. More than in any previous time, controversial speculations about the singularity, i.e. an artificial intelligence becoming self-aware and self-evolving leading to an unprecedented rapid technological change in human civilization (Eden et al. 2013), have intrigued growing interest and posed hypotheses. These attempt to define the human future and determine what it means to be human through the innovation of technologies (see Chap. 2 Risks). Most discussions of the technological singularity have emphasised a perspective on how the emergence of artificial intelligent machines outperforming human brain capability could lead to plausible extinction of the human race (Miller 2012; Kurzweil 2009; Bostrom 2002). The focus of such techno-human future visions is mainly eyed on narrow AI or AGI (Artificial General Intelligence), a single technological variable leading to an intelligence explosion. However, pre-singularity is a long-term process in which many variables can affect the timescale and sequence of pre-singularity components and thus influence the consequences of post-singularity. It is possible that technological change could be managed as a rational or planned process with a transparent agenda. We would be able to navigate the changing process and eventually create a desirable outcome of the singularity.

P. Zheng (✉)
Senior Lecturer in Business and Management, Canterbury Christ Church University, Canterbury, UK
e-mail: ping.zheng@canterbury.ac.uk

M.-A. Akhmad
Principal Scientist, Applied Intelligence Laboratories, BAE Systems plc, Chelmsford, UK
e-mail: optonet@gmail.com

V. Callaghan et al. (eds.), *The Technological Singularity*,
The Frontiers Collection, DOI 10.1007/978-3-662-54033-6_4

Technological change is an economic and social phenomenon taking place at different levels of temporal aggregation in economic and social settings, involving individual consumers, businesses, markets, science, technology, formal and informal institutions and culture at wider levels of society (Suarez-Villa 2009; Compagni et al. 2015). These variables contain a variety of technological, organisational, social, national and institutional aspects, which become agents of change that engage and interact in the technology transformation (see Fig. 1). Technological innovations in robotics, biotech, nanotech and cognitive science, and so on, can cause a paradigm shift that may subsequently affect how AI is eventually used, designed and integrated. The basic concept is summarised in Fig. 1. This chapter attempts to provide an explanation of how change agencies interact within the institutional environment in the pre-singularity process. The two primary outcomes are either a desirable 'controlled' or undesirable 'uncontrolled' singularity event arising. We also propose the condition for 'no singularity' occurring, which is explained in Sect. 4 (Anti-singularity Postulate).

4.2 Pre-singularity: The Dynamic Process of Technological Change

4.2.1 Paradigm Shifts

Paradigm Shifts are formed from a series of nudges, especially in the case of technology research, which is driven by solving present problems whilst building a foundation for future problem solving. When it comes to how we push the development of technology, how do we create paradigm shifts that benefits us all? Our fundamental way of thinking towards technological progress is a key aspect to determine a path towards a singularity event. For example, if an extra-terrestrial alien philanthropist wanted to consider investing in Earth and the potential of the human race, what would the "human race plc." sales brochure state as its mission statement?

> .....the vehement urge to survive and overcome all obstacles..........to grow our knowledge and experience through ingenious and courageous exploration of our reality.....

This fundamental motivation and philosophy of how our civilisation perpetuates forward is inexorably linked to how we embrace and utilise paradigm shifting technologies. The explosion of singularity aspects may become less likely if the developments of these technologies are carried out in-line with singularity control and monitoring mechanisms such as regulatory technology review boards (see Chap. 2 Risks 3.3.1). Relationships between the use of key technologies (in specified time frames) and possible pre-singularity scenarios can be predicted and considered. Key technology innovations and management of those technologies that will cause the pre-singularity environment can be carefully monitored and explored for possible prevention of an unexpected singularity event from occurring. Similar to 'The Observer Effect' in physics, i.e. the act of observation can change the phenomena being observed (Heisenberg 1958), the momentum of analysis and thought regarding our future and potential technological singularity is increasing and is changing how the human race will navigate its path to the singularity and beyond. The realisation of the emergence of advanced technologies and their combinations and how they could have effect on a singularity scenario are being implemented into technology roadmaps spanning the next 50+ years. Although this paradigm shift is expected, it is a spectrum of technology milestones, respective decisions from change agents and dynamic market conditions that will define the events leading to the singularity point being recognised and accepted.

4.2.2 Technological Change and Innovation Adoption

Studies of the social dynamics of technology development, including perspectives on the social construction of technology, the actor-network, human agency, have looked into the interactive effect of technological change with the socioeconomic forces (Boudreau and Robey 2005; Orlikowski 2000). Leonardi (2009) emphasises not only that social groups change a technology but also how technological and organisational changes mutually constitute one another in contexts where technology development and use occurs within a single organisation or across multiple organisations that are highly interdependent. The impact of technology can vary on different users depending on their attitude and approach of adoption. The appreciation of technology features and capability to integrate into the organisation's structure will lead to the recreation of 'sociomaterial' structure where it accommodates the use and full application of new technology. For example, Compagni et al' study of the diffusion of robotic surgery in Italy revealed that there were non-adopters (hospitals) deliberately refusing to adopt the innovation, with their justification pertaining to the uncertainty surrounding the technical benefits and economic sustainability of medical robotic systems (Compagni et al. 2015:251).

Waarts et al. (2002) point out that the most important stimulating factors for technology adopters are a combination of internal drives, such as the firm's attitude towards innovation and the strategic importance of innovation for the firm with external forces like the industry competitiveness. Early adopters in particular run a higher risk by adopting an innovative product/system (e.g. Enterprise Resource Planning) that is unproven in the long term. The ability to bear risk is a function of sufficient resources, the size and the financial strength of the company (Webster 1969; Waarts et al. 2002). The non-adopters or late adopters are more concerned with practical aspects of rapid changes in technology, costs and requisite knowledge (Robertson 1986). When businesses consider the costs and risks involved, the most common strategy is to postpone the adoption of technological innovation. Thus, new technology suppliers may possibly increase the likelihood of adoption by offering a gradual path of implementation as an alternative to the 'big-bang' approach. Apart from this, there are multiple factors affecting the diffusion and adoption of technological innovations, such as; legal regulation, institutional arrangements, government policies, market competition and organisational resources.

4.2.3 The Change Agency Perspective

The role of agency in the diversification of socioeconomic change emphasises the development of institutional environments in terms of socioeconomic frameworks of rules, routines, conventions and normative pressures (Scott 1987; Oliver 1996; Hoffman 1999). The concept of "agency" originates from the field of sociology, which has mainly focused on the interaction between human/organisational agencies and institutional orders in shaping social orders (see Lukes 1973; Coleman 1986; Emirbayer and Mische 1998; Battilana and Casciaro 2012). Actors are able to intentionally create new institutions and also effect societal-level changes (Garud et al. 2002; Seo and Creed 2002; Battilana 2006). The interplay of key actors within institutional orders is central to a better understanding of the forces of economic and social change (Lamberg and Pajunen 2010; Hodgson 2007).

To some extent, national governments can play a critical role of socioeconomic change. Take China for example; its transition to a market economy is based on the liberalisation of market relations by a major agency; the Communist Party. This has chosen to cultivate market socialism through developing a number of key agencies for social and economic transformation. It has focused upon developing technology innovation by sponsoring three types of key agency; restructured state-owned enterprises, the encouragement of foreign-owned joint ventures and the development of privately-owned entrepreneurial enterprises (Zheng and Scase 2012). The capital accumulation process as shaped by the actions and behaviour of economic agents—that is, institutions and individuals—is conditioned by a variety of socio-economic and political settings and constraints which, themselves, are the outcome of historical legacies, cultures and conflicts (North 1990; Redding and Witt 2009). Seo and Creed (2002) argue that those who arise as change agents are likely to be the groups that are dissatisfied with how their interests are accommodated within institutions and, as such, become potential forces for initiating structural change. Garud et al. (2002, 2007) also suggest that organisational fields are characterised by dominant and peripheral actors, who can act as institutional change agents because they collectively experience tensions arising from contradictions in a given socio-historical context. This body of literature focuses almost entirely on entrepreneurs as the main agent of change agents that are capable of mobilising resources to create new institutions or transforming existing ones. Other dominant economic agents in different organisational fields also impact upon structural changes and interplay within this evolving process of societal transformation. The actors of different structural environments interactively respond to the emergent problems posed by changing historical, political, social and economic situations. These orientations and actions towards emergent situations—whether they are 'deliberate' or 'improvised' (Baker et al. 2003)—are an important force of structural change. In applying the interactive role of different actors and agencies, we may gain insights into the dynamics of the 'socio-technological' change process that is interactive with and mutually conditioned by the changing forces. We highlight the discursive process that agents of change actively engage in the diffusion of innovations and social transformation.

4.2.3.1 Business Organisations as Agents of Change in Innovation Practice

Multinational and international organisations are major employers in the global market and as a result, their employment policies and practices have a strong bearing on reshaping the pool of human resources and the experience of work for international business practice. They are facilitating the reproduction of skills and knowledge required for the creation and implementation of new technologies. They are often regarded as a 'best practice' benchmark. Furthermore, they are the dominant force of innovation practices because their capabilities are based upon a combination of local and global skills and knowledge with R&D technologies and advanced management practices. They are proactive in responding to market demands because of their creative and deliberative attempts to be at the forefront of market threads. Just to name a few; Google, Apple, IBM, Samsung, Huawei are market leaders in their fields of technology innovation. Google's robot dog, Apple's intelligent Safari browser, IBM's Watson smart computer, Samsung's smart TV, Sony's dancing robots and Huawei's FusionCloud computing solution are good examples of their leading innovation products. These global giants have powerful influence over technological development. Under technology-driven capitalism, the corporate strategic focus has been shifted from production to research (Suarez-Villa 2009). Perhaps their key role as agents of change lies in their own pursuits of maintaining market leadership in innovation and isomorphic pressure on fast research outcomes. They are in the race to create new technologies, inventions and innovations against their rival forces.

4.2.3.2 Social Networks as Agents of Change

The innovation model for knowledge economies is shifting from in-house and 'closed' style to 'open' model that heavily depends on external collaboration and elusive resources (Chesbrough 2006). In the era of the Internet, organisations have become socially connected and widely embedded in an open platform within the socioeconomic structure. For example, when Apple first introduced iOS, it exercised complete control over preinstalled, native applications to ensure the value expected from the platform. Reluctantly but inevitably, Apple later embraced and integrated third party developers' creative products into its product platform that offered a sharing community introducing user-installed, unofficial applications created by individuals and businesses worldwide, which resulted in the present day Apple App Store (Yoo et al. 2012). Growing areas, such as robotics, nanotechnology, bioinformatics and so on, demand creativity more than many other resources. Their future advancement rides on the quality and capabilities of the social networks to reproduce it (Suarez-Villa. 2009).

Social media and social networks are taking precedence in product marketing communications which draw attention to the social construction processes that surround the diffusion of technological innovations. Individuals and businesses are growing more interdependent on each other through electronically connected socioeconomic infrastructure in a global context. Such external networks thus become the new resources providing for creativity and skills that innovation and entrepreneurship requires. For example, crowd-funding networks such as 'Kick-Starter' are social platforms allowing entrepreneurs or inventors to demonstrate their new ideas (mainly technology-based product ideas) to the public in an attempt at obtaining individual funds/donations for start up capital. Crowd funding has solved the seed capital problem for small businesses and creative individuals that often struggle to obtain funds from financial institutions. Crowd funding social platforms not only encourages new technology based inventions from individual actors rather than organisational actors, but also it functions as a market assessment mechanism, approving good quality proposals with market demand.

Social networks can also act as mediators. It dilutes organisational hierarchical control over the technological innovations and creates such network mechanisms as consensual decision-making, easy access, participatory discussion and open debate (Suarez-Villa 2009). These measures help sustain a network's control apparatus. In this way, networks are agents of change. The human agency, the sharing of data and processes with digital tools have altered conventional norms of ownership, roles, and rules, often triggering new configurations of relationships among actors involved in innovation networks (Yoo et al. 2012).

4.2.3.3 The Influence of Entrepreneurs as Agents of Change

Entrepreneurial firms have demonstrated considerable initiatives in pursuing and securing opportunities under institutional constraints. Private ownership is one of major significance in the development of free market economy, since it not only represents a force for contributing to the country's economic growth but also as a change agency for restructuring the economy. In a turbulent and changing marketplace, it is not surprising that many such entrepreneurs have emerged as a new class of plutocrats and business magnates. Their entrepreneurial ingenuity has led to their pursuing and securing business opportunities. They are innovative, opportunity-driven and risk-taking. Regardless of a lack of resources, they develop tactics and networks to gain access to required skills and capital. There is a growing number of technology driven entrepreneurs and series entrepreneurs who are taking the lead role in technology innovations. They are extremely adaptive and flexible; their pursuit for new opportunities and collective actions for achieving their own interests have created unintended consequences for the technological and socioeconomic change. As such, they are key agents of technological change and societal transformation.

Take Chinese SME, Alibaba for example; it was regarded as the most valuable IPO in human history, valued at 45 billion USD in the New York Stock market. As Jack Ma, founder of Alibaba addressed in one of his public speeches in the US, he created Alibaba with the intention to help Chinese SMEs to access overseas markets which he not only influenced the way SMEs internationalised and developed but also had led key inventions in internet communication technologies to help secure online payment e.g. Zhifubao (Chinese version of Paypal system) and face recognition software for online payment. Another example of an influential entrepreneur is Elon Musk, a co-founder of and CEO of Tesla Motors. He has pushed forward the technological development of capable electric vehicle technology. His visions for sustainable technologies are influencing investment in alternative and disruptive technology research, which include; the SpaceX (advancing the state of rocket technology), the Hyperloop transportation system, SolarCity's photovoltaic technology and now his aspiration of helping humanity to colonise Mars (Coppinger 2016). In the new era of rapid technological change, ethical and responsible entrepreneurship is important. After the 2009 financial crisis triggered in America and extended to the UK and EU countries, there had been many calls from politicians for 'responsible capitalism' in the United Kingdom (Rae 2015). Clearly, the pursuit of sole profit-driven goal has led the banking crisis and market failure. Entrepreneurship under malicious profit maximisation has become destructive in value creation. Entrepreneurs as the important agents of change are expected to work in socially responsible and environmentally sustainable ways. This need becomes more crucial in the new tech era where there are increasing involvements of technology-based entrepreneurial ventures as the driving force for innovations.

4.2.3.4 Nation States as Agents of Change

Often state-owned or state-sponsored organisations are highly policy-led and characterised as reactive to restructuring demands as they are inclined to conform with institutional arrangements. Their resource-focused competitive advantage is built upon on a grant-aided policy (Hassard et al. 2010). They can enjoy privileged access to state resources and have favourable policy lobby in foreign markets. For example, in China, mergers and acquisitions are taking place between medium to large state organisations across sectors, which enable diverse state-owned businesses to agglomerate to form larger industrial clusters. Such a strategy is creating quasi-monopoly industries that dominate vital sectors of the Chinese economy, such as, for example, telecom and real estate sectors (Zheng and Scase 2012). Nation states can play a critical role in shaping the technological development in state-owned organisations and institutes. According to OECD (2012), the country' spending on R&D has showed steady increase, US at 2.7 percent, Japan at 3.2 percent and China is about 1.6 percent. It is estimated that China's double-digit growth in R&D spending is expected to match and surpass that of the US by about 2023, as R&D growth over the past 15 years has consistently exceeded 10 percent (R&D Magazine 2011:2). In June 2013, China's supercomputer—'Tianhe-2' (developed by the National University of Defence Technology)—became the world's most powerful computing system, superseding the current US system. This is part of the government's effort to make the country's high-tech industries more competitive and less dependent on overseas rivals. It is evident that Chinese government has made significant improvement in intellectual property laws and indigenous innovation policy. Its continuous restructuring of free market and socioeconomic environments yield rapid growth of both private economy and state sectors. These purposive actions and interplay of change agents have also created those superpower companies (e.g. China Petro, China Telecom, Huawei and Haier) who have leading technological innovations in their fields and have influential power over domestic international marketplace. In China, owing to direct state control over the actions of state sectors that dominate large segments of the economy, the Chinese governments can quickly intervene or order them to ramp up investments and leading to leverage economic growth and market change (Zheng and Scase 2013). This type of interdependence will continue to shape the evolution of technological change process in China.

4.3 Key Drivers of Technology Research and Their Impact

Although there are a plethora of potential technologies that could contribute to a singularity event, there are key future technologies which are predicted to make a direct impact to a path towards the singularity. These include the exact science of human brain functions, quantum computing, advanced global communication networks, artificial intelligence programming, nanotechnology, new methods for energy generation and specific defence technologies. The quest to fully understand and replicate human brain functions is on going. Incremental approaches to understand and simulate brain functionality have been based around analysing individual synapses and neurons and manufacturing equivalent artificial mechanisms to simulate a very small portion of the potential parallel computing capability. This method of development will allow this technology to scale up further, which will lead to larger simulated neural networks to be formed, paving the way to a complete simulation of human brain processing. One such current project looking at exactly this is the Human Brain Project, and international collaborative programme funded by the European Commission (The European Commission 2015).

From a general research and development point of view the complete understanding of the human brain is an important component underpinning a singularity event. Not only does the goal of understanding the human brain stem from technology development to explore the human condition from a biological perspective, it is an important milestone to reverse engineer for computer science. As the human brain is currently considered as the most advanced computer in existence, it is only logical to reverse engineer, modify and improve on what has naturally evolved to develop more advanced computing technology. Although this particular development path is not the only way toward computers with human capability equivalence (such as traditional electronics design and quantum computing), when it is successful it will: (1) be a significant marker event to a singularity when human brain function is simulated on an artificial computing platform and (2) definitively prove that compatibilities exist between artificial computing technologies and human brain functions meaning that augmented human brain technology can be realised (an aspect of transhumanism).

Quantum technology research has been gaining momentum over the last ten years and is now dubbed as "quantum 2.0", as quantum physics is now being applied directly to technological innovation. This includes, quantum computing, quantum based sensors, imaging and cryptography. As future quantum technology may be viewed as 'disruptive', i.e. it may create new product markets that may displace similar technology with lesser capability. It has been an area of investment that has been increasing. For example, in 2014 the UK Government has announced £270 million funding to assist the transition of UK quantum research into the commercial sector (Physics World, p.7 Feb 2014). Quantum 2.0 technology is important, as it is likely to be the foundation of technologies that will cause an evolution of a singularity scenario. Quantum computing will provide processing platforms that are orders of magnitude more efficiency that current systems, allowing for computers with higher capacities than the human brain (Kurzweil 2009). Quantum research and development will also perpetuate the general miniaturisation of technology and the need to develop systems to cope with extremely larges amounts of information.

Research into artificial intelligence, is also a fundamental technology component for a singularity to occur, especially in the case of a future AI/AGI establishing itself. At present, AI developments are characterised into specific areas including, machine vision, perception, logical decision-making and problem solving, motion and manipulation. Slowly but surely, the automated systems that we directly and indirectly come across are becoming more powerful, with more control over outputs.

The rise of 'Cyber' technologies is also an important factor to the future singularity. Specifically, the rise of cyber defence research and development and the range of technologies it impacts. Over the last decade there has been a surge of investment in this area (both in the public and private sector) primarily due to the exponential growth and usage of the Internet. Of course, cyber defence/security is quite broad and ranges from an individual protecting their information from being used and manipulated to overt and covert military based cyber attacks between countries. However, as technology is intertwined with this issue, a computer or mobile device connected to the Internet for the individual to advanced computing systems controlling critical infrastructure (such as power generation, telecommunications and transportation) the potential cyber threats to these systems spur research and development into protecting against such advanced threats. There is increased demand for the use of more powerful and intelligent computing platforms to thwart of threats faster. Larger and more efficient information management systems are required to understand cyber threats and assist in decision-making. Advanced, semi-autonomous computer programs designed to simulate, detect and defend against attacks on critical systems. It is possible that the very motivation to develop these technologies (in this case to keep the technological advantage, among others) will significantly contribute to a singularity occurring.

One perspective of Cyber defence/security technology providing a good foundation for a singularity type event is with the perpetuation of an AGI. All major governments have some sort of Cyber defence organisation. All of which invest in the latest computing technology and have extensive budgets to maintain and upgrade their capability to maintain an advantage. This is the same with large corporations such as IBM, and Apple, although the emphasis would be to protect their corporate information and the manipulation thereof, rather than to prevent national cyber crimes. This technology comprises of the latest processing power, advanced computer programs, vast data management and storage capabilities, all of which are key for an AGI to perpetuate itself, if one is born in the future. At the very least the Cyber defence industry is significantly pushing forward the development of important technologies for a singularity event to occur.

4.4 The Anti-singularity Postulate

As previously mentioned 'The Observer Effect' in quantum physics also applies to the possibility of an anti-singularity arising. That is a series of events which either leads to (a) the expected or random selection of future technologies not converging to produce a singularity; (b) the technologies to create an singularity exist but are controlled to ensure a singularity event does not occur; or (c) stagnant development of relevant technologies due to externalities (i.e. economic crisis, war or natural disasters).

The fact that the above paragraph has been written means that it reinforces other contemplations about the potential threats from a singularity event, specifically the rise of an artificial intelligence that is uncontrolled and could negatively impact the human race on a whim. Therefore, the stated and recorded thoughts regarding a singularity event from the introduction of Asimov's laws of robotics in 1942 to the present will inherently change the course of current and future relevant technology development to ensure that it does not happen at all. Or if the environment of the singularity does arise, it is controlled and monitored to ensure a 'big bang' event does not occur. This also implies that there is a certain amount of human 'common sense' embedded into infinite variables that make up part of the 'existence probability of a singularity' equation.

The potential negative impacts of a singularity has generated (at least for now) soft guidelines and policies regarding monitoring and development of key technologies to an extent. Soft artificial intelligence is being continually developed, to improve the autonomy of smart devices and systems and to generally improve the reliability of intelligent hardware and software that the human race is dependent upon to preserve our way of life. One example of this is Apple's Siri and Microsoft's Cortana voice activated personal assistants on smart mobile devices. They are certainly intelligent enough to understand what the user is saying and to respond with the appropriate action accurately as well as having the ability to continually learn from past action to improve its own service to the end user. However, there are safeguards in place to switch off this intelligent system for people who do not want their privacy to affected. i.e. the collation of the end user's personal behaviour and actions to improve the service of these quasi AI interfaces. Privacy guidelines and policies relating to soft AI interfaces may ensure that there is always a 'hard on/off' switch in-built into these systems, which ultimately mean that they can be reset, controlled, monitored and less susceptible to contribute to a singularity event (see Chap. 2 Risks, 4.1.3).

Another technology area which may directly contribute to an anti-singularity are Cyber defence technologies. For most large corporations and governments, cyber defence mechanisms will be in place to prevent things such as hacking and computer virus infection. Due to the fundamental nature these protection measures, it may limit the effects of AI trying to establish itself in a potential singularity event from trying to utilise networked computing resources. One example of this is the extensive use of software and hardware encryption. If we assume that an AI becomes conscious and decides to replicate and control significant computer resources, it will still have to learn how to overcome and bypass many different types of encryption, providing enough time for automated defence systems and to be made aware of its breach attempts and to affect an appropriate response. That in itself poses and interesting point; how can a potential hard AI singularity threat (in its initial stages) attempting to take control possibly win against a plethora of quasi AI cyber defence programs? All of these programs are specifically designed and honed to stop unwanted breaches of its systems. The likelihood is that it would fail in the initial stages of trying to establish itself and quickly highlight itself to being a threat to be dealt with.

There is also the possibility of stagnant development of key technologies causing a singularity event to occur. This could be because of a natural disaster (earth quake, tsunami etc.) meaning the technology development shifts to remedy and prevent damage from further disasters or man-made disasters such as wars. Of course there is significant evidence showing that leaps in technology were apparent during wartime, such as the atomic bomb in World War Two. However, since war causes enemies to constantly counter each other's developments (technological or otherwise) whilst battling for control of resources. This also enforces the argument that an optimal environment for a destructive singularity may not exist in this context.

4.5 Conclusions

Institutions reflect the order and conventions of a society that is often defined by law and regulations (Dobbin 2004). Thus, governments as agencies represent institutions that prescribes certain practices while circumscribing others. Industrial organisations play collective actions to set up norms and standards of practice to stabilise their position in the marketplace. Entrepreneurs and powerful individuals often play active role within the existing institutions with deliberate or improvised attempts to promote new institutional arrangements that meet their needs and interests. The institutional change process is dynamic and interactive which drives the evolutional advancement of society. Technological and cultural development will inevitably trigger institutional changes, as incompatibilities between agencies and institutional structures arise. Incompatibilities and threats arising from current control and regulatory systems will encourage the creation of adapted and new institutions that can co-evolve with new technological norms and demands.

We argue that change agency theory can incorporate a variety of technological developments, where new innovation and their related risks provide not a barrier but an opportunity for new institutional environments to occur. Further exploration of the interactive effect between change agents and institutional and social contexts is required to understand how AI and human agency relationships could be institutionalised and legitimated to provide a positive human/AGI coexistence. To do so, it would be necessary for governments to develop 'AI Institutions' where a systematic arrangement of institutions can be established to regularise and structure the co-existence of AGI and human race. AI Institutions may include the following aspects:

(1) Policing mechanisms to use AI to regulate AGI.
(2) Economic policies focusing on the role of AGI in general economic development.
(3) Financial policies targeted on the role and function of AGI in the capital market.
(4) Social networks mechanisms to monitor the future role, societal impact and moral hazards of AGI applications.
(5) 'Think Tank' organisations such as the Oxford University Future of Humanity Institute continually identifying and understanding the evolving cultural, moral and cognitive norms of AGI in society (http://www.fhi.ox.ac.uk/).

(6) Global collaboration between governments to build transparency in AGI developments and an effective worldwide AI monitoring scheme. This may be aided by the general increase in globalisation.

It is important to recognise that, in the revolutionary process leading to a potential singularity, the change agencies such as individuals, organisations and governments play key roles in determining not only how innovations are carried out and applied but also what institutional change may occur to regularise the technological development. The co-evolution of the human race and technology is generally an iterative process, providing the opportunity to predict implement structural and institutional control mechanisms to minimise the risk of an unwanted singularity event. Possible future prevention measures may also incorporate counter productive change agencies such as non-cooperative governments and organisations that may try to circumvent AGI development regulation.

We need to develop our awareness and measures against potential dangers and be proactive to make prevention mechanisms. We need to have confidence and faith in mankind and believe that AI will continue to be assistance to humans.

References

Baker, T.; Miner, A. S. and Eesley, D. T. (2003), 'Improvising Firms: Briolage, Account Giving and Improvisational Competencies in the Founding Process', *Research Policy*, 32, 255–76.

Battilana, J. (2006), 'Agency and Institutions: The Enabling Role of Individuals' Social Position', *Organisation*, 13 (5), 653–76.

Battilana, J. and Casciaro, T. (2012), 'Change agents, networks, and institutions: a contingency theory of organisational change', *Academy of Management Journal*, 55 (2), 381–398.

Bostrom, N. (2002), 'Existential risks: analysing human extinction scenarios and related harzards', *Journal of Evolution and Technology*, 9.

Boudreau, M. C. and Robey, D. (2005), 'Enacting integrated information technology: a human agency perspective', *Organization Science*, 16 (1), 3–18.

Chesbrough H. (2006), *Open Business Models: How to Thrive in the New Innovation Landscape*. Boston, MA: Harvard Business School Press.

Coleman, J. S. (1986), 'Social theory, social research, and a theory of action', *American Journal of Sociology*, 6, 1309–35.

Coppinger, R. (2016), 'Elon Musk outlines Mars colony vision', Science & Environment, 27th September 2016, *BBC News*. (http://www.bbc.co.uk/news/science-environment-37486372 accessed on 29th September 2016)

Compagni, A., Mele, V. and Ravasi, D. (2015), 'How early implementations influence later adoptions of innovation: social positioning and skill reproduction in the diffusion of robotic surgery', *Academy of Management Journal*, 58 (1), 242–278.

Dobbin, F. (2004), The New Economic Sociology. Princeton, New York: Princeton University Press.

Eden, A. H., Moor, J. H., Soraker, J. H., Steinhart, E. (2013), 'Singularity Hypotheses: A Scientific and Philosophical Assessment', Berlin: Springer.

Emirbayer, M. and Mische, A. (1998), 'What is Agency?', *American Journal of Sociology*, 103 (4), 962–1023.

Garud, R.; Hardy, C. and Maguire, S. (2007), 'Institutional entrepreneurship as embedded agency: an introduction to the special issue', *Organisation Studies*, 28, 957–969.

Garud, R.; Jain, S.; and Kumaraswamy, A. (2002), 'Institutional entrepreneurship in the sponsorship of common technological standards: the Case of sun Microsystems and Java', *Academy of Management Journal,* 45 (1), 196–214.

Hassard, J., Morris, J., Sheehan, J. and Xiao, J. (2010), 'China's state-owned enterprises: economic reform and organisational restructuring', *Journal of Organisational Change Management,* 23 (5), 500–516.

Heisenberg, W. (1958), *Physics and Philosophy: the revolution in modern science*, London: Unwin University Books.

Hodgson, G. M. (2007), 'Institutions and individuals: interaction and evolution', *Organisation Studies,* 28 (1), 95–116.

Hoffman, A. J. (1999), 'Institutional evolution and change: environmentalism and the US chemical industry', *Academy of Management Journal*, 42, 351–371.

Kurzweil, R. (2009), *The singularity is near*, 2nd ed. London: Gerald Duckworth.

Lamberg, J. A. and Pajunen, K. (2010), 'Agency, institutional change, and continuity: The case of the Finish civil war', *Journal of Management Studies*, 47 (5), 815–36.

Leonardi, P. M. (2009), 'Organising technology: toward a theory of sociomaterial imbrication', *Academy of Management Annual Meeting Proceedings.*

Lukes, S. (1973), *Individualism*, Oxford: Basil Blackwell.

Miller, J. D. (2012), 'Some economic incentives facing a business that might bring about a technological singularity', chapter 8 in "Singularity Hypotheses, The Frontiers Collection" (eds) by Eden et al., Berlin: Springer.

North, D. C. (1990), *Institutions, institutional change and economic performance*, Cambridge: Cambridge University Press.

OECD (Organisation for Economic Co-operation and Development) (2012), 'Active with the People's Republic of China', pp. 1–61.

Oliver, C. (1996), 'The institutional embeddedness of economic activity', *Advances in Strategic Management*, 13, 163–186.

Orlikowski, W. J. (2000), 'Using technology and constituting structures: a practice lens for studying technology in organisations', *Organisation Science*, 11 (4), 404–428.

Physic World (2014), February Issue, p 2.

R&D Magazine (2011), 'China's R&D Momentum: 2012 Global R&D Funding Forecasting', *R&D Magazine*, 53 (7), 60–61.

Rae, D. (2015), *Opportunity-centred Entrepreneurship*, 2nd ed. London: Palgrave.

Redding, G. and Witt, M. A. (2009), 'China's business system and its future trajectory', *Asia Pacific Journal of Management*, September, 26 (3), 381–399.

Robertson, T. S. (1986), 'Competitive effects on technology diffusion', *Journal of Marketing*, 50 (1), 1–12.

Scott, W. R. (1987), 'The adolesence of institutional theory', *Administrative Science Quarterly*, 32 (4), 493.

Seo, M-G. and Creed, W. E. D. (2002), 'Institutional Contradictions, Praxis, and Institutional Change: A Dialectical Perspective', *Academy of Management Review*, 27 (2), 222–247.

Suarez-Villa, L. (2009), *Technocapitalism*, Philadephia: Temple University Press.

The European Commission http://www.humanbrainproject.eu (accessed on 1st March 2015).

Waarts, E., Yvonne, M. v. E. and Hillegersberg, J. v. (2002), 'The dynamics of factors affecting the adoption of innovations', *The Journal of Product Innovation Management*, 19, 412–423.

Webster, F. E. (1969), 'New Product adoption in industrial markets: a framework for analysis', *Journal of Marketing*, 33 (1), 35–39.

Yoo, Y., Boland Jr, R. J., Lyytinen, K. and Majchrzak, A. (2012), 'Organising for innovation in the digitalized world', *Organisation Science*, 23 (5), 1398–1408.

Zheng, P. and Scase, R. (2012), 'The restructuring of market socialism: the contribution of 'Agency' theoretical perspective', *Thunderbird International Business Review*, 55 (1), 103–114.

Zheng, P. and Scase, R. (2013), *Emerging business ventures under market socialism: entrepreneurship in China*, Routledge Studies in International Business and The World Economy, London: Routledge.

Chapter 5
Agent Foundations for Aligning Machine Intelligence with Human Interests: A Technical Research Agenda

Nate Soares and Benya Fallenstein

5.1 Introduction

The property that has given humans a dominant advantage over other species is not strength or speed, but intelligence. If progress in artificial intelligence continues unabated, AI systems will eventually exceed humans in general reasoning ability. A system that is "superintelligent" in the sense of being "smarter than the best human brains in practically every field" could have an enormous impact upon humanity (Bostrom 2014). Just as human intelligence has allowed us to develop tools and strategies for controlling our environment, a superintelligent system would likely be capable of developing its own tools and strategies for exerting control (Muehlhauser and Salamon 2012). In light of this potential, it is essential to use caution when developing AI systems that can exceed human levels of general intelligence, or that can facilitate the creation of such systems.

Since artificial agents would not share our evolutionary history, there is no reason to expect them to be driven by human motivations such as lust for power. However, nearly all goals can be better met with more resources (Omohundro 2008). This suggests that, by default, superintelligent agents would have incentives to acquire resources currently being used by humanity. (Just as artificial agents would not automatically acquire a lust for power, they would not automatically acquire a human sense of fairness, compassion, or conservatism.) Thus, most goals would put the agent at odds with human interests, giving it incentives to deceive or manipulate its human operators and resist interventions designed to change or debug its behavior (Bostrom 2014, Chap. 8).

N. Soares (✉) · B. Fallenstein
Machine Intelligence Research Institute, Berkeley, USA
e-mail: nate@intelligence.org

B. Fallenstein
e-mail: benya@intelligence.org

V. Callaghan et al. (eds.), *The Technological Singularity*,
The Frontiers Collection, DOI 10.1007/978-3-662-54033-6_5

Care must be taken to avoid constructing systems that exhibit this default behavior. In order to ensure that the development of smarter-than-human intelligence has a positive impact on the world, we must meet three formidable challenges: How can we create an agent that will reliably pursue the goals it is given? How can we formally specify beneficial goals? And how can we ensure that this agent will assist and cooperate with its programmers as they improve its design, given that mistakes in early AI systems are inevitable?

This agenda discusses technical research that is tractable today, which the authors think will make it easier to confront these three challenges in the future. Sections 5.2 through 5.4 motivate and discuss six research topics that we think are relevant to these challenges.

We call a smarter-than-human system that reliably pursues beneficial goals "aligned with human interests" or simply "aligned."[1] To become confident that an agent is aligned in this way, a practical implementation that merely *seems* to meet the challenges outlined above will not suffice. It is also important to gain a solid formal understanding of why that confidence is justified. This technical agenda argues that there is foundational research we can make progress on today that will make it easier to develop aligned systems in the future, and describes ongoing work on some of these problems.

Of the three challenges, the one giving rise to the largest number of currently tractable research questions is the challenge of finding an agent architecture that will reliably and autonomously pursue a set of objectives—that is, an architecture that can at least be aligned with *some* end goal. This requires theoretical knowledge of how to design agents which reason well and behave as intended even in situations never envisioned by the programmers. The problem of highly reliable agent designs is discussed in Sect. 5.2.

The challenge of developing agent designs which are tolerant of human error also gives rise to a number of tractable problems. We expect that smarter-than-human systems would by default have incentives to manipulate and deceive human operators, and that special care must be taken to develop agent architectures which avert these incentives and are otherwise tolerant of programmer error. This problem and some related open questions are discussed in Sect. 5.3.

Reliable and error-tolerant agent designs are only beneficial if the resulting agent actually pursues desirable outcomes. The difficulty of concretely specifying what is meant by "beneficial behavior" implies a need for some way to construct agents that reliably *learn* what to value (Bostrom 2014, Chap. 12). A solution to this "value learning" problem is vital; attempts to start making progress are reviewed in Sect. 5.4.

[1]A more careful wording might be "aligned with the interests of sentient beings." We would not want to benefit humans at the expense of sentient non-human animals—or (if we build them) at the expense of sentient machines.

Why work on these problems now, if smarter-than-human AI is likely to be decades away? This question is touched upon briefly below, and is discussed further in Sect. 5.5. In short, the authors believe that there are theoretical prerequisites for designing aligned smarter-than-human systems over and above what is required to design misaligned systems. We believe that research can be done today that will make it easier to address alignment concerns in the future.

5.1.1 Why These Problems?

This technical agenda primarily covers topics that the authors believe are *tractable*, *uncrowded*, *focused*, and *unable to be outsourced* to forerunners of the target AI system.

By *tractable* problems, we mean open problems that are concrete and admit immediate progress. Significant effort will ultimately be required to align real smarter-than-human systems with beneficial values, but in the absence of working designs for smarter-than-human systems, it is difficult if not impossible to begin most of that work in advance. This agenda focuses on research that can help us gain a better understanding today of the problems faced by almost any sufficiently advanced AI system. Whether practical smarter-than-human systems arise in ten years or in one hundred years, we expect to be better able to design safe systems if we understand solutions to these problems.

This agenda further limits attention to *uncrowded* domains, where there is not already an abundance of research being done, and where the problems may not be solved over the course of "normal" AI research. For example, program verification techniques are absolutely crucial in the design of extremely reliable programs (Sotala and Yampolskiy 2015, Sect. 5.5, this volume), but program verification is not covered in this agenda primarily because a vibrant community is already actively studying the topic.

This agenda also restricts consideration to *focused* tools, ones that would be useful for designing aligned systems in particular (as opposed to intelligent systems in general). It might be possible to design generally intelligent AI systems before developing an understanding of highly reliable reasoning sufficient for constructing an aligned system. This could lead to a risky situation where powerful AI systems are built long before the tools needed to safely utilize them. Currently, significant research effort is focused on improving the capabilities of artificially intelligent systems, and comparatively little effort is focused on superintelligence alignment (Bostrom 2014, Chap. 14). For that reason, this agenda focuses on research that improves our ability to design aligned systems in particular.

Lastly, we focus on research that *cannot be safely delegated to machines*. As AI algorithms come to rival humans in scientific inference and planning, new possibilities will emerge for outsourcing computer science labor to AI algorithms themselves.

This is a consequence of the fact that *intelligence* is the technology we are designing: on the path to great intelligence, much of the work may be done by smarter-than-human systems.[2]

As a result, the topics discussed in this agenda are ones that we believe are difficult to safely delegate to AI systems. Error-tolerant agent design is a good example: no AI problem (including the problem of error-tolerant agent design itself) can be safely delegated to a highly intelligent artificial agent that has incentives to manipulate or deceive its programmers. By contrast, a sufficiently capable automated engineer would be able to make robust contributions to computer vision or natural language processing even if its own visual or linguistic abilities were initially lacking. Most intelligent agents optimizing for some goal would also have incentives to improve their visual and linguistic abilities so as to enhance their ability to model and interact with the world.

It would be risky to delegate a crucial task before attaining a solid theoretical understanding of exactly what task is being delegated. It may be possible to use our understanding of ideal Bayesian inference to task a highly intelligent system with developing increasingly effective approximations of a Bayesian reasoner, but it would be far more difficult to delegate the task of "finding good ways to revise how confident you are about claims" to an intelligent system *before* gaining a solid understanding of probability theory. The theoretical understanding is useful to ensure that the right questions are being asked.

5.2 Highly Reliable Agent Designs

Bird and Layzell (2002) describe a genetic algorithm which, tasked with making an oscillator, re-purposed the printed circuit board tracks on the motherboard as a makeshift radio to amplify oscillating signals from nearby computers. This is not kind of solution the algorithm would have found if it had been simulated on a virtual circuit board possessing only the features that *seemed* relevant to the problem. Intelligent search processes in the real world have the ability to use resources in unexpected ways, e.g., by finding "shortcuts" or "cheats" not accounted for in a simplified model.

When constructing intelligent systems which learn and interact with all the complexities of reality, it is not sufficient to verify that the algorithm behaves well in test settings. Additional work is necessary to verify that the system will continue working as intended in application. This is especially true of systems possessing general intelligence at or above the human level: superintelligent machines might find

[2]Since the Dartmouth Proposal (McCarthy et al. 1955), it has been a standard idea in AI that a sufficiently smart machine intelligence could be intelligent enough to improve itself. In 1965, I.J. Good observed that this might create a positive feedback loop leading to an "intelligence explosion" (Good 1965). Sotala and Yampolskiy (2015, Sect. 2.3, this volume) and Bostrom (2014, Chap. 14) has observed that an intelligence explosion is especially likely if the agent has the ability to acquire more hardware, improve its software, or design new hardware.

strategies and execute plans beyond both the experience and imagination of the programmers, making the clever oscillator of Bird and Layzell look trite. At the same time, unpredictable behavior from smarter-than-human systems could cause catastrophic damage, if they are not aligned with human interests (Yudkowsky 2008).

Because the stakes are so high, testing combined with a gut-level intuition that the system will continue to work outside the test environment is insufficient, even if the testing is extensive. It is important to also have a *formal* understanding of precisely why the system is expected to behave well in application.

What constitutes a formal understanding? It seems essential to us to have both (1) an understanding of precisely what problem the system is intended to solve; and (2) an understanding of precisely why *this* practical system is expected to solve *that* abstract problem. The latter must wait for the development of practical smarter-than-human systems, but the former is a theoretical research problem that we can already examine.

A full description of the problem would reveal the conceptual tools needed to understand why practical heuristics are expected to work. By analogy, consider the game of chess. Before designing practical chess algorithms, it is necessary to possess not only a predicate describing checkmate, but also a description of the problem in term of trees and backtracking algorithms: Trees and backtracking do not immediately yield a practical solution—building a full game tree is infeasible—but they are the conceptual tools of computer chess. It would be quite difficult to justify confidence in a chess heuristic before understanding trees and backtracking.

While these conceptual tools may seem obvious in hindsight, they were not clear to foresight. Consider the famous essay by Edgar Allen Poe about Maelzel's Mechanical Turk (Poe 1836). It is in many ways remarkably sophisticated: Poe compares the Turk to "the calculating machine of Mr. Babbage" and then remarks on how the two systems cannot be of the same kind, since in Babbage's algebraical problems each step follows of necessity, and so can be represented by mechanical gears making deterministic motions; while in a chess game, no move follows with necessity from the position of the board, and even if our own move followed with necessity, the opponent's would not. And so (argues Poe) we can see that chess cannot possibly be played by mere mechanisms, only by thinking beings. From Poe's state of knowledge, Shannon's (1950) description of an idealized solution in terms of backtracking and trees constitutes a great insight. Our task it to put theoretical foundations under the field of general intelligence, in the same sense that Shannon put theoretical foundations under the field of computer chess.

It is possible that these foundations will be developed over time, during the normal course of AI research: in the past, theory has often preceded application. But the converse is also true: in many cases, application has preceded theory. The claim of this technical agenda is that, in safety-critical applications where mistakes can put lives at risk, it is crucial that certain theoretical insights come first.

A smarter-than-human agent would be embedded within and computed by a complex universe, learning about its environment and bringing about desirable states of affairs. How is this formalized? What metric captures the question of how well an agent would perform in the real world?[3]

Not all parts of the problem must be solved in advance: the task of designing smarter, safer, more reliable systems could be delegated to early smarter-than-human systems, if the research done by those early systems can be sufficiently trusted. It is important, then, to focus research efforts particularly on parts of the problem where an increased understanding is necessary to construct a minimal reliable generally intelligent system. Moreover, it is important to focus on aspects which are currently tractable, so that progress can in fact be made today, and on issues relevant to alignment in particular, which would not otherwise be studied over the course of "normal" AI research.

In this section, we discuss four candidate topics meeting these criteria: (1) *realistic world-models*, the study of agents learning and pursuing goals while embedded within a physical world; (2) *decision theory*, the study of idealized decision-making procedures; (3) *logical uncertainty*, the study of reliable reasoning with bounded deductive capabilities; and (4) *Vingean reflection*, the study of reliable methods for reasoning about agents that are more intelligent than the reasoner. We will now discuss each of these topics in turn.

5.2.1 Realistic World-Models

Formalizing the problem of computer intelligence may seem easy in theory: encode some set of preferences as a utility function, and evaluate the expected utility that would be obtained if the agent were implemented. However, this is not a full specification: What is the set of "possible realities" used to model the world? Against what distribution over world models is the agent evaluated? How is a given world model used to score an agent? To ensure that an agent would work well in reality, it is first useful to formalize the problem faced by agents learning (and acting in) arbitrary environments.

Solomonoff (1964) made an early attempt to tackle these questions by specifying an "induction problem" in which an agent must construct world models and promote correct hypotheses based on the observation of an arbitrarily complex environment, in a manner reminiscent of scientific induction. In this problem, the agent and environment are separate. The agent gets to see one bit from the environment in each turn, and must predict the bits which follow.

Solomonoff's induction problem answers all three of the above questions in a simplified setting: The set of world models is any computable environment

[3]Legg and Hutter (2007) provide a preliminary answer to this question, by defining a "universal measure of intelligence" which scores how well an agent can learn the features of an external environment and maximize a reward function. This is the type of formalization we are looking for: a scoring metric which describes how well an agent would achieve some set of goals. However, while the Legg-Hutter metric is insightful, it makes a number of simplifying assumptions, and many difficult open questions remain (Soares 2015).

(e.g., any Turing machine). In reality, the simplest hypothesis that predicts the data is generally correct, so agents are evaluated against a simplicity distribution. Agents are scored according to their ability to predict their next observation. These answers were insightful, and led to the development of many useful tools, including algorithmic probability and Kolmogorov complexity.

However, Solomonoff's induction problem does not fully capture the problem faced by an agent learning about an environment while embedded *within* it, as a subprocess. It assumes that the agent and environment are separated, save only for the observation channel. What is the analog of Solomonoff induction for agents that are embedded within their environment?

This is the question of *naturalized induction* (Bensinger 2013). Unfortunately, the insights of Solomonoff do not apply in the naturalized setting. In Solomonoff's setting, where the agent and environment are separated, one can consider arbitrary Turing machines to be "possible environments." But when the agent is embedded in the environment, consideration must be restricted to environments which embed the agent. Given an algorithm, what is the set of environments which embed that algorithm? Given that set, what is the analogue of a simplicity prior which captures the fact that simpler hypotheses are more often correct?

Technical problem (Naturalized Induction) *What, formally, is the induction problem faced by an intelligent agent embedded within and computed by its environment? What is the set of environments which embed the agent? What constitutes a simplicity prior over that set? How is the agent scored? For discussion, see Soares (2015).*

Just as a formal description of Solomonoff induction answered the above three questions in the context of an agent learning an external environment, a formal description of naturalized induction may well yield answers to those questions in the context where agents are embedded in and computed by their environment.

Of course, the problem of computer intelligence is not simply an induction problem: the agent must also interact with the environment. Hutter (2000) extends Solomonoff's induction problem to an "interaction problem," in which an agent must both learn and act upon its environment. In each turn, the agent both observes one input and writes one output, and the output affects the behavior of the environment. In this problem, the agent is evaluated in terms of its ability to maximize a reward function specified in terms of inputs. While this model does not capture the difficulties faced by agents which are embedded within their environment, it does capture a large portion of the problem faced by agents interacting with arbitrarily complex environments. Indeed, the interaction problem (and AIXI Hutter 2000, its solution) are the basis for the "universal measure of intelligence" developed by Legg and Hutter (2007).

However, even barring problems arising from the agent/environment separation, the Legg-Hutter metric does not fully characterize the problem of computer intelligence. It scores agents according to their ability to maximize a reward function specified in terms of observation. Agents scoring well by the Legg-Hutter metric are extremely effective at ensuring their observations optimize a reward function, but these high-scoring agents are likely to be the type that find clever ways to seize control of their observation channel rather than the type that identify and manipulate the features in the world that the reward function was intended to proxy for (Soares 2015). Reinforcement learning techniques which punish the agent for attempting to take control would only incentivize the agent to deceive and mollify the programmers until it found a way to gain a decisive advantage (Bostrom 2014, Chap. 8).

The Legg-Hutter metric does not characterize the question of how well an algorithm would perform if implemented in reality: to formalize that question, a scoring metric must evaluate the resulting environment histories, not just the agent's observations (Soares 2015).

But human goals are not specified in terms of environment histories, either: they are specified in terms of high-level notions such as "money" or "flourishing humans." Leaving aside problems of philosophy, imagine rating a system according to how well it achieves a straightforward, concrete goal, such as by rating how much diamond is in an environment after the agent has acted on it, where "diamond" is specified concretely in terms of an atomic structure. Now the goals are specified in terms of atoms, and the environment histories are specified in terms of Turing machines paired with an interaction history. How is the environment history evaluated in terms of atoms? This is the *ontology identification* problem.

Technical problem (Ontology Identification) *Given goals specified in some ontology and a world model, how can the ontology of the goals be identified in the world model? What types of world models are amenable to ontology identification? For a discussion, see Soares (2015).*

To evaluate world models, the world models must be evaluated in terms of the ontology of the goals. This may be difficult in cases where the ontology of the goals does not match reality: it is one thing to locate atoms in a Turing machine using an atomic model of physics, but it is another thing altogether to locate atoms in a Turing machine modeling quantum physics. De Blanc (2011) further motivates the idea that explicit mechanisms are needed to deal with changes in the ontology of the system's world model.

Agents built to solve the wrong problem—such as optimizing their observations—may well be capable of attaining superintelligence, but it is unlikely that those agents could be aligned with human interests (Bostrom 2014, Chap. 12). A better understanding of naturalized induction and ontology identification is needed to fully specify the problem that intelligent agents would face when pursuing human goals while embedded within reality, and this increased understanding could be a crucial tool when it comes to designing highly reliable agents.

5.2.2 *Decision Theory*

Smarter-than-human systems must be trusted to make good decisions, but what does it mean for a decision to be "good"? Formally, given a description of an environment and an agent embedded within, how is the "best available action" identified, with respect to some set of preferences? This is the question of decision theory.

The answer may seem trivial, at least in theory: simply iterate over the agent's available actions, evaluate what would happen if the agent took that action, and then return whichever action leads to the most expected utility. But this is not a full specification: How are the "available actions" identified? How is what "would happen" defined?

The difficulty is easiest to illustrate in a deterministic setting. Consider a deterministic agent embedded in a deterministic environment. There is exactly one action that the agent will take. Given a set of actions that it "could take," it is necessary to evaluate, for each action, what would happen if the agent took that action. But the agent will not take most of those actions. How is the counterfactual environment constructed, in which a deterministic algorithm "does something" that, in the real environment, it doesn't do? Answering this question requires a theory of counterfactual reasoning, and counterfactual reasoning is not well understood.

Technical problem (Theory of Counterfactuals) *What theory of counterfactual reasoning can be used to specify a procedure which always identifies the best action available to a given agent in a given environment, with respect to a given set of preferences? For discussion, see Soares and Fallenstein (2014).*

Decision theory has been studied extensively by philosophers. The study goes back to Pascal, and has been picked up in modern times by Lehmann (1950), Wald (1939), Jeffrey (1983), Joyce (1999), Lewis (1981), Pearl (2000), and many others. However, no satisfactory method of counterfactual reasoning yet answers this particular question. To give an example of why counterfactual reasoning can be difficult, consider a deterministic agent playing against a perfect copy of itself in the classic prisoner's dilemma (Rapoport and Chammah 1965). The opponent is guaranteed to do the same thing as the agent, but the agents are "causally separated," in that the action of one cannot physically affect the action of the other.

What is the counterfactual world in which the agent on the left cooperates? It is not sufficient to consider changing the action of the agent on the left while holding the action of the agent on the right constant, because while the two are causally disconnected, they are logically constrained to behave identically. Standard causal reasoning, which neglects these logical constraints, misidentifies "defection" as the best strategy available to each agent even when they know they have identical source codes (Lewis 1979).[4] Satisfactory counterfactual reasoning must respect these logical constraints, but how are logical constraints formalized and identified? It is fine to say that, in the counterfactual where the agent on the left cooperates, all identical copies of it also cooperate; but what counts as an identical copy? What if the right

[4]As this is a multi-agent scenario, the problem of counterfactuals can also be thought of as game-theoretic. The goal is to define a procedure which reliably identifies the best available action; the label of "decision theory" is secondary. This goal subsumes both game theory and decision theory: the desired procedure must identify the best action in all settings, even when there is no clear demarcation between "agent" and "environment." Game theory informs, but does not define, this area of research.

agent runs the same algorithm written in a different programming language? What if it only does the same thing 98% of the time?

These questions are pertinent in reality: practical agents must be able to identify good actions in settings where other actors base their actions on imperfect (but well-informed) predictions of what the agent will do. Identifying the best action available to an agent requires taking the non-causal logical constraints into account. A satisfactory formalization of counterfactual reasoning requires the ability to answer questions about how other deterministic algorithms behave in the counterfactual world where the agent's deterministic algorithm does something it doesn't. However, the evaluation of "logical counterfactuals" is not yet well understood.

Technical problem (Logical Counterfactuals) *Consider a counterfactual in which a given deterministic decision procedure selects a different action from the one it selects in reality. What are the implications of this counterfactual on other algorithms? Can logical counterfactuals be formalized in a satisfactory way? A method for reasoning about logical counterfactuals seems necessary in order to formalize a more general theory of counterfactuals. For a discussion, see Soares and Fallenstein (2014).*

Unsatisfactory methods of counterfactual reasoning (such as the causal reasoning of Pearl (2000)) seem powerful enough to support smarter-than-human intelligent systems, but systems using those reasoning methods could systematically act in undesirable ways (even if otherwise aligned with human interests).

To construct practical heuristics that are known to make good decisions, even when acting beyond the oversight and control of humans, it is essential to understand what is meant by "good decisions." This requires a formulation which, given a description of an environment, an agent embedded in that environment, and some set of preferences, identifies the best action available to the agent. While modern methods of counterfactual reasoning do not yet allow for the specification of such a formula, recent research has pointed the way towards some promising paths for future research.

For example, Wei Dai's "updateless decision theory" (UDT) is a new take on decision theory that systematically outperforms causal decision theory (Hintze 2014), and two of the insights behind UDT highlight a number of tractable open problems (Soares and Fallenstein 2014).

Recently, Barasz et al. (2014) developed a concrete model, together with a Haskell implementation, of multi-agent games where agents have access to each others' source code and base their decisions on what they can prove about their opponent. They have found that it is possible for some agents to achieve robust cooperation in the one-shot prisoner's dilemma while remaining unexploitable Barasz et al. (2014).

These results suggest a number of new ways to approach the problem of counterfactual reasoning, and we are optimistic that continued study will prove fruitful.

5.2.3 Logical Uncertainty

Consider a reasoner encountering a black box with one input chute and two output chutes. Inside the box is a complex Rube Goldberg machine that takes an input ball and deposits it in one of the two output chutes. A probabilistic reasoner may have uncertainty about where the ball will exit, due to uncertainty about which Rube Goldberg machine is in the box. However, standard probability theory assumes that if the reasoner *did* know which machine the box implemented, they would know where the ball would exit: the reasoner is assumed to be *logically omniscient*, i.e., to know all logical consequences of any hypothesis they entertain.

By contrast, a practical bounded reasoner may be able to know exactly which Rube Goldberg machine the box implements without knowing where the ball will come out, due to the complexity of the machine. This reasoner is *logically uncertain.* Almost all practical reasoning is done under some form of logical uncertainty (Gaifman 2004), and almost all reasoning done by a smarter-than-human agent must be some form of logically uncertain reasoning. Any time an agent reasons about the consequences of a plan, the effects of running a piece of software, or the implications of an observation, it must do some sort of reasoning under logical uncertainty. Indeed, the problem of an agent reasoning about an environment in which it is embedded as a subprocess is inherently a problem of reasoning under logical uncertainty.

Thus, to construct a highly reliable smarter-than-human system, it is vitally important to ensure that the agent's logically uncertain reasoning is reliable and trustworthy. This requires a better understanding of the theoretical underpinnings of logical uncertainty, to more fully characterize what it means for logically uncertain reasoning to be "reliable and trustworthy" (Soares and Fallenstein 2015).

It is natural to consider extending standard probability theory to include the consideration of worlds which are "logically impossible" (e.g., where a deterministic Rube Goldberg machine behaves in a way that it doesn't). This gives rise to two questions: What, precisely, are logically impossible possibilities? And, given some means of reasoning about impossible possibilities, what is a reasonable prior probability distribution over impossible possibilities?

The problem is difficult to approach in full generality, but a study of logical uncertainty in the restricted context of assigning probabilities to logical sentences goes back at least to Łoś (1955) and Gaifman (1964), and has been investigated by many, including Halpern (2003), Hutter et al. (2013), Demski (2012), Russell (2014), and others. Though it isn't clear to what degree this formalism captures the kind of logically uncertain reasoning a realistic agent would use, logical sentences in, for example, the language of Peano Arithmetic are quite expressive: for example, given the Rube Goldberg machine discussed above, it is possible to form a sentence which is true if and only if the machine deposits the ball into the top chute. Thus, considering reasoners which are uncertain about logical sentences is a useful starting point. The problem of assigning probabilities to sentences of logic naturally divides itself into two parts.

First, how can probabilities consistently be assigned to sentences? An agent assigning probability 1 to short contradictions is hardly reasoning about the sentences as if they are logical sentences: some of the logical structure must be preserved. But which aspects of the logical structure? Preserving all logical implications requires that the reasoner be deductively omnipotent, as some implications $\phi \rightarrow \psi$ may be very involved. The standard answer in the literature is that a coherent assignment of probabilities to sentences corresponds to a probability distribution over complete, consistent logical theories (Gaifman 1964; Christiano 2014a); that is, an "impossible possibility" is any consistent assignment of truth values to all sentences. Deductively limited reasoners cannot have fully coherent distributions, but they can approximate these distributions: for a deductively limited reasoner, "impossible possibilities" can be any assignment of truth values to sentences that looks consistent so far, so long as the assignment is discarded as soon as a contradiction is introduced.

Technical problem (Impossible Possibilities) *How can deductively limited reasoners approximate reasoning according to a probability distribution over complete theories of logic? For a discussion, see Christiano (2014a).*

Second, what is a satisfactory prior probability distribution over logical sentences? If the system is intended to reason according to a theory at least as powerful as Peano Arithmetic (PA), then that theory will be incomplete (Gödel et al. 1934). What prior distribution places nonzero probability on the set of complete extensions of PA? Deductively limited agents would not be able to literally use such a prior, but if it were computably approximable, then they could start with a rough approximation of the prior and refine it over time. Indeed, the process of refining a logical prior—getting better and better probability estimates for given logical sentences—captures the whole problem of reasoning under logical uncertainty in miniature. Hutter et al. (2013) have defined a desirable prior, but Sawin and Demski (2013) have shown that it cannot be computably approximated. Demski (2012) and Christiano (2014a) have also proposed logical priors, but neither seems fully satisfactory. The specification of satisfactory logical priors is difficult in part because it is not yet clear which properties are desirable in a logical prior, nor which properties are possible.

Technical problem (Logical Priors) *What is a satisfactory set of priors over logical sentences that a bounded reasoner can approximate? For a discussion, see Soares and Fallenstein (2015).*

Many existing tools for studying reasoning, such as game theory, standard probability theory, and Bayesian networks, all assume that reasoners are logically omniscient. A theory of reasoning under logical uncertainty seems necessary to formalize the problem of naturalized induction, and to generate a satisfactory theory of counterfactual reasoning. If these tools are to be developed, extended, or improved, then a better understanding of logically uncertain reasoning is required.

5.2.4 Vingean Reflection

Instead of specifying superintelligent systems directly, it seems likely that humans may instead specify generally intelligent systems that go on to create or attain superintelligence. In this case, the reliability of the resulting superintelligent system depends upon the reasoning of the initial system which created it (either anew or via self-modification).

If the agent reasons reliably under logical uncertainty, then it may have a generic ability to evaluate various plans and strategies, only selecting those which seem beneficial. However, some scenarios put that logically uncertain reasoning to the test more than others. There is a qualitative difference between reasoning about simple programs and reasoning about human-level intelligent systems. For example, modern program verification techniques could be used to ensure that a "smart" military drone obeys certain rules of engagement, but it would be a different problem altogether to verify the behavior of an artificial military general which must run an entire war. A general has far more autonomy, ability to come up with clever unexpected strategies, and opportunities to impact the future than a drone.

A self-modifying agent (or any that constructs new agents more intelligent than itself) must reason about the behavior of a system that is more intelligent than the reasoner. This type of reasoning is critically important to the design of self-improving agents: if a system will attain superintelligence through self-modification, then the impact of the system depends entirely upon the correctness of the original agent's reasoning about its self-modifications (Fallenstein and Soares 2015).

Before trusting a system to attain superintelligence, it seems prudent to ensure that the agent uses appropriate caution when reasoning about successor agents.[5] That is, it seems necessary to understand the mechanisms by which agents reason about smarter systems.

Naive tools for reasoning about plans including smarter agents, such as backwards induction (Ben-Porath 1997), would have the reasoner evaluate the smarter agent by simply checking what the smarter agent would do. This does not capture the difficulty of the problem: a parent agent cannot simply check what its successor agent would do in all scenarios, for if it could, then it would already know what actions its successor will take, and the successor would not in any way be smarter.

Yudkowsky and Herreshoff (2013) call this observation the "Vingean principle," after Vernor Vinge (1993), who emphasized how difficult it is for humans to predict the behavior of smarter-than-human agents. Any agent reasoning about more intelligent successor agents must do so *abstractly*, without pre-computing all actions that the successor would take in every scenario. We refer to this kind of reasoning as *Vingean reflection*.

[5]Of course, if an agent reasons *perfectly* under logical uncertainty, it would also reason well about the construction of successor agents. However, given the fallibility of human reasoning and the fact that this path is critically important, it seems prudent to verify the agent's reasoning methods in this scenario specifically.

Technical problem (Vingean Reflection) *How can agents reliably reason about agents which are smarter than themselves, without violating the Vingean principle? For discussion, see Fallenstein and Soares (2015).*

It may seem premature to worry about how agents reason about self-improvements before developing a theoretical understanding of reasoning under logical uncertainty in general. However, it seems to us that work in this area can inform understanding of what sort of logically uncertain reasoning is necessary to reliably handle Vingean reflection.

Given the high stakes when constructing systems smarter than themselves, artificial agents might use some form of extremely high-confidence reasoning to verify the safety of potentially dangerous self-modifications. When *humans* desire extremely high reliability, as is the case for (e.g.) autopilot software, we often use formal logical systems (United States Department of Defense 1985; United Kingdom Ministry of Defense 1991). High-confidence reasoning in critical situations may require something akin to formal verification (even if *most* reasoning is done using more generic logically uncertain reasoning), and so studying Vingean reflection in the domain of formal logic seems like a good starting point.

Logical models of agents reasoning about agents that are "more intelligent," however, run into a number of obstacles. By Gödel's second incompleteness theorem (1934), sufficiently powerful formal systems cannot rule out the possibility that they may be inconsistent. This makes it difficult for agents using formal logical reasoning to verify the reasoning of similar agents which also use formal logic for high-confidence reasoning; the first agent cannot verify that the latter agent is consistent. Roughly, it seems desirable to be able to develop agents which reason as follows:

> This smarter successor agent uses reasoning similar to mine, and my own reasoning is sound, so its reasoning is sound as well.

However, Gödel et al. (1934) showed that this sort of reasoning leads to inconsistency, and these problems do in fact make Vingean reflection difficult in a logical setting (Fallenstein and Soares 2015; Yudkowsky 2013).

Technical problem (Löbian Obstacle) *How can agents gain very high confidence in agents that use similar reasoning systems, while avoiding paradoxes of self-reference? For discussion, see Fallenstein and Soares (2015).*

These results may seem like artifacts of models rooted in formal logic, and may seem irrelevant given that practical agents must eventually use logical uncertainty rather than formal logic to reason about smarter successors. However, it has been shown that many of the Gödelian obstacles carry over into early probabilistic logics in a straightforward way, and some results have already been shown to apply in the domain of logical uncertainty (Fallenstein 2014).

Studying toy models in this formal logical setting has led to partial solutions (Fallenstein and Soares 2014). Recent work has pushed these models towards probabilistic settings (Fallenstein and Soares 2014; Yudkowsky 2014; Soares 2014). Further research may continue driving the development of methods for reasoning under logical uncertainty which can handle Vingean reflection in a reliable way (Fallenstein and Soares 2015).

5.3 Error-Tolerant Agent Designs

Incorrectly specified superintelligent agents could be dangerous (Yudkowsky 2008). Correcting a modern AI system involves simply shutting the system down and modifying its source code. Modifying a smarter-than-human system may prove more difficult: a system attaining superintelligence could acquire new hardware, alter its software, create subagents, and take other actions that would leave the original programmers with only dubious control over the agent. This is especially true if the agent has incentives to resist modification or shutdown. If intelligent systems are to be safe, they must be constructed in such a way that they are amenable to correction, even if they have the ability to prevent or avoid correction.

This does not come for free: by default, agents have incentives to preserve their own preferences, even if those conflict with the intentions of the programmers (Omohundro 2008; Soares and Fallenstein 2015). Special care is needed to specify agents that avoid the default incentives to manipulate and deceive (Bostrom 2014, Chap. 8).

Restricting the actions available to a superintelligent agent may be quite difficult (Bostrom 2014, Chap. 9). Intelligent optimization processes often find unexpected ways to fulfill their optimization criterion using whatever resources are at their disposal; recall the evolved oscillator of Bird and Layzell (2002). Superintelligent optimization processes may well use hardware, software, and other resources in unanticipated ways, making them difficult to contain if they have incentives to escape.

We must learn how to design agents which do not have incentives to escape, manipulate, or deceive in the first place: agents which reason as if they are incomplete and potentially flawed in dangerous ways, and which are therefore amenable to online correction. Reasoning of this form is known as "corrigible reasoning."

Technical problem (Corrigibility) *What sort of reasoning can reflect the fact that an agent is incomplete and potentially flawed in dangerous ways? For discussion, see Soares and Fallenstein (2015).*

Naïve attempts at specifying corrigible behavior are unsatisfactory. For example, "moral uncertainty" frameworks could allow agents to learn values through observation and interaction, but would still incentivize agents to resist changes to the underlying moral uncertainty framework if it happened to be flawed. Simple "penalty terms" for manipulation and deception also seem doomed to failure: agents subject to such penalties would have incentives to resist modification while cleverly avoiding the technical definitions of "manipulation" and "deception." The goal is not to design systems that fail in their attempts to deceive the programmers; the goal is to construct reasoning methods that do not give rise to deception incentives in the first place.

A formalization of the intuitive notion of corrigibility remains elusive. Active research is currently focused on small toy problems, in the hopes that insight gained there will generalize. One such toy problem is the "shutdown problem," which involves designing a set of preferences that incentivize an agent to shut down upon the press of a button without also incentivizing the agent to either cause or prevent the pressing of that button (Soares and Fallenstein 2015). Stuart Armstrong's utility indifference technique (2015) provides a partial solution, but not a fully satisfactory one.

Technical problem (Utility Indifference) *Can a utility function be specified such that agents maximizing that utility function switch their preferences on demand, without having incentives to cause or prevent the switching? For discussion, see Armstrong (2015).*

A better understanding of corrigible reasoning is essential to design agent architectures that are tolerant of human error. Other research could also prove fruitful, including research into reliable containment mechanisms. Alternatively, agent designs could somehow incentivize the agent to have a "low impact" on the world. Specifying "low impact" is trickier than it may seem: How do you tell an agent that it can't affect the physical world, given that its RAM is physical? How do you tell it that it can only use its own hardware, without allowing it to use its motherboard as a makeshift radio? How do you tell it not to cause big changes in the world when its behavior influences the actions of the programmers, who influence the world in chaotic ways?

Technical problem (Domesticity) *How can an intelligent agent be safely incentivized to have a low impact? Specifying such a thing is not as easy as it seems. For a discussion, see Armstrong et al. (2012).*

Regardless of the methodology used, it is crucial to understand designs for agents that could be updated and modified during the development process, so as to ensure that the inevitable human errors do not lead to catastrophe.

5.4 Value Specification

A highly-reliable, error-tolerant agent design does not guarantee a positive impact; the effects of the system still depend upon whether it is pursuing appropriate goals.

A superintelligent system may find clever, unintended ways to achieve the specific goals that it is given. Imagine a superintelligent system designed to cure cancer which does so by stealing resources, proliferating robotic laboratories at the expense of the biosphere, and kidnapping test subjects: the intended goal may have been "cure cancer without doing anything bad," but such a goal is rooted in cultural context and shared human knowledge.

It is not sufficient to construct systems that are smart enough to figure out the intended goals. Human beings, upon learning that natural selection "intended" sex to be pleasurable only for purposes of reproduction, do not suddenly decide that contraceptives are abhorrent. While one should not anthropomorphize natural selection, humans are capable of understanding the process which created them while being completely unmotivated to alter their preferences. For similar reasons, when developing AI systems, is not sufficient to develop a system intelligent enough to figure out the intended goals; the system must also somehow be deliberately constructed to pursue them (Bostrom 2014, Chap. 8).

However, the "intentions" of the operators are a complex, vague, fuzzy, context-dependent notion (Yudkowsky 2011; cf. Sotala and Yampolskiy 2015, Sects. 2.2 and 5.2.5, this volume). Concretely writing out the full intentions of the operators in a machine-readable format is implausible if not impossible, even for simple tasks. An intelligent agent must be designed to learn and act according to the preferences of its operators.[6] This is the *value learning problem.*

Directly programming a rule which identifies cats in images is implausibly difficult, but specifying a system that inductively learns how to identify cats in images is possible. Similarly, while directly programming a rule capturing complex human intentions is implausibly difficult, intelligent agents could be constructed to inductively learn values from training data.

Inductive value learning presents unique difficulties. The goal is to develop a system which can classify potential outcomes according to their value, but what sort of training data allows this classification? The labeled data could be given in terms of the agent's world-model, but this is a brittle solution if the ontology of the world-model is liable to change. Alternatively, the labeled data could come in terms of observations, in which case the agent would have to first learn how the labels in the observations map onto objects in the world-model, and *then* learn how to classify outcomes. Designing algorithms which can do this likely requires a better understanding of methods for constructing multi-level world-models from sense data.

Technical problem (Multi-Level World-Models) *How can multi-level world-models be constructed from sense data in a manner amenable to ontology identification? For a discussion, see Soares (2016).*

Standard problems of inductive learning arise, as well: how could a training set be constructed which allows the agent to fully learn the complexities of value? It is easy to imagine a training set which labels many observations of happy humans as "good" and many observations of needlessly suffering humans as "bad," but the simplest generalization from this data set may well be that humans value human-shaped things

[6]Or of all humans, or of all sapient creatures, etc. There are many philosophical concerns surrounding what sort of goals are ethical when aligning a superintelligent system, but a solution to the value learning problem will be a practical necessity regardless of which philosophical view is the correct one.

mimicking happy emotions: after training on this data, an agent may be inclined to construct many simple animatronics mimicking happiness. Creating a training set that covers all relevant dimensions of human value is difficult for the same reason that specifying human value directly is difficult. In order for inductive value learning to succeed, it is necessary to construct a system which identifies ambiguities in the training set—dimensions along which the training set gives no information—and queries the operators accordingly.

Technical problem (Ambiguity Identification) *Given a training data set and a world model, how can dimensions which are neglected by the training data be identified? For discussion, see Soares (2016).*

This problem is not unique to value learning, but it is especially important for it. Research into the programmatic identification of ambiguities, and the generation of "queries" which are similar to previous training data but differ along the ambiguous axis, would assist in the development of systems which can safely perform inductive value learning.

Intuitively, an intelligent agent should be able to use some of its intelligence to assist in this process: it does not take a detailed understanding of the human psyche to deduce that humans care more about some ambiguities (are the human-shaped things actually humans?) than others (does it matter if there is a breeze?). To build a system that acts as intended, the system must model the intentions of the operators and act accordingly. This adds another layer of indirection: the system must model the operators in some way, and must extract "preferences" from the operator-model and update its preferences accordingly (in a manner robust against improvements to the model of the operator). Techniques such as inverse reinforcement learning (Ng and Russell 2000), in which the agent assumes that the operator is maximizing some reward function specified in terms of observations, are a good start, but many questions remain unanswered.

Technical problem (Operator Modeling) *By what methods can an operator be modeled in such a way that (1) a model of the operator's preferences can be extracted; and (2) the model may eventually become arbitrarily accurate and represent the operator as a subsystem embedded within the larger world? For a discussion, see Soares (2016).*

A system which acts as the operators intend may still have significant difficulty answering questions that the operators themselves cannot answer: imagine humans trying to design an artificial agent to do what they would want, if they were better people. How can normative uncertainty (uncertainty about moral claims) be resolved? Bostrom (2014, Chap. 13) suggests an additional level of indirection: task the system with reasoning about what sorts of conclusions the operators would come to if they had more information and more time to think. Formalizing this is difficult, and the problems are largely still in the realm of philosophy rather than technical research. However, Christiano (2014b) has sketched one possible method by which the volition of a human could be extrapolated, and Soares (2016) discusses some potential pitfalls.

Philosophical problem (Normative Uncertainty) *What ought one do when one is uncertain about what one ought to do? What norms govern uncertainty about normative claims? For a discussion, see MacAskill (2014).*

Human operators with total control over a superintelligent system could give rise to a moral hazard of extraordinary proportions by putting unprecedented power into the hands of a small few (Bostrom 2014, Chap. 6). The extraordinary potential of superintelligence gives rise to many ethical questions. When constructing autonomous agents that will have a dominant ability to determine the future, it is important to design the agents to not only act according to the wishes of the operators, but also in others' common interest. Here we largely leave the philosophical questions aside, and remark only that those who design systems intended to surpass human intelligence will take on a responsibility of unprecedented scale.

5.5 Discussion

Sections 5.2 through 5.4 discussed six research topics where the authors think that further research could make it easier to develop aligned systems in the future. This section discusses reasons why we think useful progress can be made today.

5.5.1 Toward a Formal Understanding of the Problem

Are the problems discussed above tractable, uncrowded, focused, and unlikely to be solved automatically in the course of developing increasingly intelligent AI systems?

They are certainly not very crowded. They also appear amenable to progress in the near future, though it is less clear whether they can be fully solved.

When it comes to focus, some think that problems of decision theory and logical uncertainty sound more like generic theoretical AI research than alignment-specific research. A more intuitive set of alignment problems might put greater emphasis on AI constraint (see Chap. 4 in this book) or value learning.

Progress on the topics outlined in this agenda might indeed make it easier to design intelligent systems in general. Just as the intelligence metric of Legg and Hutter (2007) lent insight into the ideal priors for agents facing Hutter's interaction problem, a full description of the naturalized induction problem could lend insight into the ideal priors for agents embedded within their universe. A satisfactory theory of logical uncertainty could lend insight into general intelligence more broadly. An ideal decision theory could reveal an ideal decision-making procedure for real agents to approximate.

But while these advancements may provide tools useful for designing intelligent systems in general, they would make it markedly easier to design aligned systems in particular. Developing a general theory of highly reliable decision-making, even if it is too idealized to be directly implemented, gives us the conceptual tools needed to design and evaluate safe heuristic approaches. Conversely, if we must evaluate real systems composed of practical heuristics before formalizing the theoretical problems that those heuristics are supposed to solve, then we will be forced to rely on our intuitions.

This theoretical understanding might not be developed by default. Causal counterfactual reasoning, despite being suboptimal, might be good enough to enable the construction of a smarter-than-human system. Systems built from poorly understood heuristics might be capable of creating or attaining superintelligence for reasons we don't quite understand—but it is unlikely that such systems could then be aligned with human interests.

Sometimes theory precedes application, but sometimes it does not. The goal of much of the research outlined in this agenda is to ensure, in the domain of superintelligence alignment—where the stakes are incredibly high—that theoretical understanding comes first.

5.5.2 *Why Start Now?*

It may seem premature to tackle the problem of AI goal alignment now, with superintelligent systems still firmly in the domain of futurism. However, the authors think it is important to develop a formal understanding of AI alignment well in advance of making design decisions about smarter-than-human systems. By beginning our work early, we inevitably face the risk that it may turn out to be irrelevant; yet failing to make preparations at all poses substantially larger risks.

We have identified a number of unanswered foundational questions relating to the development of general intelligence, and at present it seems possible to make some promising inroads. We think that the most responsible course, then, is to begin as soon as possible.

Weld and Etzioni (1994) directed a "call to arms" to computer scientists, noting that "society will reject autonomous agents unless we have some credible means of making them safe." We are concerned with the opposite problem: what if society fails to reject systems that are unsafe? What will be the consequences if someone believes a smarter-than-human system is aligned with human interests when it is not?

This is our call to arms: regardless of whether research efforts follow the path laid out in this document, significant effort must be focused on the study of superintelligence alignment as soon as possible.

References

Armstrong S (2015) AI motivated value selection, accepted to the 1st International Workshop on AI and Ethics, held within the 29th AAAI Conference on Artificial Intelligence (AAAI-2015), Austin, TX

Armstrong S, Sandberg A, Bostrom N (2012) Thinking inside the box: Controlling and using an oracle AI. Minds and Machines 22(4):299–324

Bárász M, Christiano P, Fallenstein B, Herreshoff M, LaVictoire P, Yudkowsky E (2014) Robust cooperation in the Prisoner's Dilemma: Program equilibrium via provability logic, unpublished manuscript. Available via arXiv. http://arxiv.org/abs/1401.5577

Ben-Porath E (1997) Rationality, Nash equilibrium, and backwards induction in perfect-information games. Review of Economic Studies 64(1):23–46

Bensinger R (2013) Building phenomenological bridges. Less Wrong Blog http://lesswrong.com/lw/jd9/building_phenomenological_bridges/

Bird J, Layzell P (2002) The evolved radio and its implications for modelling the evolution of novel sensors. In: Proceedings of the 2002 Congress on Evolutionary Computation. Vol. 2, IEEE, Honolulu, HI, pp 1836–1841

Bostrom N (2014) Superintelligence: Paths, Dangers, Strategies. Oxford University Press, New York

Christiano P (2014a) Non-omniscience, probabilistic inference, and metamathematics. Tech. Rep. 2014–3, Machine Intelligence Research Institute, Berkeley, CA, http://intelligence.org/files/Non-Omniscience.pdf

Christiano P (2014b) Specifying "enlightened judgment" precisely (reprise). Ordinary Ideas Blog http://ordinaryideas.wordpress.com/2014/08/27/specifying-enlightened-judgment-precisely-reprise/

de Blanc P (2011) Ontological crises in artificial agents' value systems. Tech. rep., The Singularity Institute, San Francisco, CA, http://arxiv.org/abs/1105.3821

Demski A (2012) Logical prior probability. In: Bach J, Goertzel B, Iklé M (eds) Artificial General Intelligence, Springer, New York, 7716, pp 50–59, 5th International Conference, AGI 2012, Oxford, UK, December 8–11, 2012. Proceedings

Fallenstein B (2014) Procrastination in probabilistic logic. Working paper, Machine Intelligence Research Institute, Berkeley, CA, http://intelligence.org/files/ProbabilisticLogicProcrastinates.pdf

Fallenstein B, Soares N (2014) Problems of self-reference in self-improving space-time embedded intelligence. In: Goertzel B, Orseau L, Snaider J (eds) Artificial General Intelligence, Springer, New York, 8598, pp 21–32, 7th International Conference, AGI 2014, Quebec City, QC, Canada, August 1–4, 2014. Proceedings

Fallenstein B, Soares N (2015) Vingean reflection: Reliable reasoning for self-improving agents. Tech. Rep. 2015–2, Machine Intelligence Research Institute, Berkeley, CA, https://intelligence.org/files/VingeanReflection.pdf

Gaifman H (1964) Concerning measures in first order calculi. Israel Journal of Mathematics 2(1): 1–18

Gaifman H (2004) Reasoning with limited resources and assigning probabilities to arithmetical statements. Synthese 140(1–2):97–119

Gödel K, Kleene SC, Rosser JB (1934) On Undecidable Propositions of Formal Mathematical Systems. Institute for Advanced Study, Princeton, NJ

Good IJ (1965) Speculations concerning the first ultraintelligent machine. In: Alt FL, Rubinoff M (eds) Advances in Computers, vol 6, Academic Press, New York, pp 31–88

Halpern JY (2003) Reasoning about Uncertainty. MIT Press, Cambridge, MA

Hintze D (2014) Problem class dominance in predictive dilemmas. Tech. rep., Machine Intelligence Research Institute, Berkeley, CA, http://intelligence.org/files/ProblemClassDominance.pdf

Hutter M (2000) A theory of universal artificial intelligence based on algorithmic complexity, unpublished manuscript. Available via arXiv. http://arxiv.org/abs/cs/0004001

Hutter M, Lloyd JW, Ng KS, Uther WTB (2013) Probabilities on sentences in an expressive logic. Journal of Applied Logic 11(4):386–420

Jeffrey RC (1983) The Logic of Decision, 2nd edn. Chicago University Press, Chicago, IL

Joyce JM (1999) The Foundations of Causal Decision Theory. Cambridge Studies in Probability, Induction and Decision Theory, Cambridge University Press, New York, NY

Legg S, Hutter M (2007) Universal intelligence: A definition of machine intelligence. Minds and Machines 17(4):391–444

Lehmann EL (1950) Some principles of the theory of testing hypotheses. Annals of Mathematical Statistics 21(1):1–26

Lewis D (1979) Prisoners' dilemma is a Newcomb problem. Philosophy & Public Affairs 8(3): 235–240, http://www.jstor.org/stable/2265034

Lewis D (1981) Causal decision theory. Australasian Journal of Philosophy 59(1):5–30

Łoś J (1955) On the axiomatic treatment of probability. Colloquium Mathematicae 3(2):125–137, http://eudml.org/doc/209996

MacAskill W (2014) Normative uncertainty. PhD thesis, St Anne's College, University of Oxford, http://ora.ox.ac.uk/objects/uuid:8a8b60af-47cd-4abc-9d29-400136c89c0f

McCarthy J, Minsky M, Rochester N, Shannon C (1955) A proposal for the Dartmouth summer research project on artificial intelligence. Proposal, Formal Reasoning Group, Stanford University, Stanford, CA

Muehlhauser L, Salamon A (2012) Intelligence explosion: Evidence and import. In: Eden A, Søraker J, Moor JH, Steinhart E (eds) Singularity Hypotheses: A Scientific and Philosophical Assessment, Springer, Berlin, the Frontiers Collection

Ng AY, Russell SJ (2000) Algorithms for inverse reinforcement learning. In: Langley P (ed) Proceedings of the Seventeenth International Conference on Machine Learning (ICML-'00), Morgan Kaufmann, San Francisco, pp 663–670

Omohundro SM (2008) The basic AI drives. In: Wang P, Goertzel B, Franklin S (eds) Artificial General Intelligence 2008, IOS, Amsterdam, no. 171 in Frontiers in Artificial Intelligence and Applications, pp 483–492, proceedings of the First AGI Conference

Pearl J (2000) Causality: Models, Reasoning, and Inference, 1st edn. Cambridge University Press, New York, NY

Poe EA (1836) Maelzel's chess-player. Southern Literary Messenger 2(5):318–326

Rapoport A, Chammah AM (1965) Prisoner's Dilemma: A Study in Conflict and Cooperation, Ann Arbor Paperbacks, vol 165. University of Michigan Press, Ann Arbor, MI

Russell S (2014) Unifying logic and probability: A new dawn for AI? In: Information Processing and Management of Uncertainty in Knowledge-Based Systems: 15th International Conference, IPMU 2014, Montpellier, France, July 15–19, 2014, Proceedings, Part I, Springer, no. 442 in Communications in Computer and Information Science, pp 10–14

Sawin W, Demski A (2013) Computable probability distributions which converge on π_1 will disbelieve true π_2 sentences. Tech. rep., Machine Intelligence Research Institute, Berkeley, CA, http://intelligence.org/files/Pi1Pi2Problem.pdf

Shannon CE (1950) XXII. Programming a computer for playing chess. Philosophical Magazine 41(314):256–275

Soares N (2014) Tiling agents in causal graphs. Tech. Rep. 2014–5, Machine Intelligence Research Institute, Berkeley, CA, http://intelligence.org/files/TilingAgentsCausalGraphs.pdf

Soares N (2015) Formalizing two problems of realistic world-models. Tech. Rep. 2015–3, Machine Intelligence Research Institute, Berkeley, CA, https://intelligence.org/files/RealisticWorldModels.pdf

Soares N (2016) The value learning problem. In: Ethics for Artificial Intelligence Workshop at the 25th International Joint Conference on Artificial Intelligence (IJCAI-16). New York, NY, July 9th-15th

Soares N, Fallenstein B (2014) Toward idealized decision theory. Tech. Rep. 2014–7, Machine Intelligence Research Institute, Berkeley, CA, https://intelligence.org/files/TowardIdealizedDecisionTheory.pdf

Soares N, Fallenstein B (2015) Questions of reasoning under logical uncertainty. Tech. Rep. 2015–1, Machine Intelligence Research Institute, Berkeley, CA, https://intelligence.org/files/QuestionsLogicalUncertainty.pdf

Solomonoff RJ (1964) A formal theory of inductive inference. Part I. Information and Control 7(1):1–22

United Kingdom Ministry of Defense (1991) Requirements for the procurement of safety critical software in defence equipment. Interim Defence Standard 00-55, United Kingdom Ministry of Defense

United States Department of Defense (1985) Department of Defense trusted computer system evaluation criteria. Department of Defense Standard DOD 5200.28-STD, United States Department of Defense, http://csrc.nist.gov/publications/history/dod85.pdf

Vinge V (1993) The coming technological singularity: How to survive in the post-human era. In: Vision-21: Interdisciplinary Science and Engineering in the Era of Cyberspace, NASA Lewis Research Center, no. 10129 in NASA Conference Publication, pp 11–22, http://ntrs.nasa.gov/archive/nasa/casi.ntrs.nasa.gov/19940022856.pdf

Wald A (1939) Contributions to the theory of statistical estimation and testing hypotheses. Annals of Mathematical Statistics 10(4):299–326

Weld D, Etzioni O (1994) The first law of robotics (a call to arms). In: Hayes-Roth B, Korf RE (eds) Proceedings of the Twelfth National Conference on Artificial Intelligence, AAAI Press, Menlo Park, CA, pp 1042–1047, http://www.aaai.org/Papers/AAAI/1994/AAAI94-160.pdf

Yudkowsky E (2008) Artificial intelligence as a positive and negative factor in global risk. In: Bostrom N, Ćirković MM (eds) Global Catastrophic Risks, Oxford University Press, New York, pp 308–345

Yudkowsky E (2011) Complex value systems in Friendly AI. In: Schmidhuber J, Thórisson KR, Looks M (eds) Artificial General Intelligence, Springer, Berlin, no. 6830 in Lecture Notes in Computer Science, pp 388–393, 4th International Conference, AGI 2011, Mountain View, CA, USA, August 3–6, 2011. Proceedings

Yudkowsky E (2013) The procrastination paradox. Brief technical note, Machine Intelligence Research Institute, Berkeley, CA, http://intelligence.org/files/ProcrastinationParadox.pdf

Yudkowsky E (2014) Distributions allowing tiling of staged subjective EU maximizers. Tech. rep., Machine Intelligence Research Institute, Berkeley, CA, http://intelligence.org/files/DistributionsAllowingTiling.pdf

Yudkowsky E, Herreshoff M (2013) Tiling agents for self-modifying AI, and the Löbian obstacle. Early draft, Machine Intelligence Research Institute, Berkeley, CA, http://intelligence.org/files/TilingAgents.pdf

Chapter 6
Risk Analysis and Risk Management for the Artificial Superintelligence Research and Development Process

Anthony M. Barrett and Seth D. Baum

6.1 Introduction

A substantial amount of work has made the case that global catastrophic risks (GCRs) deserve special attention (Sagan 1983; Ng 1991; Bostrom 2002; Beckstead 2013; Maher and Baum 2013). Major issues in addressing GCRs include assessing the probabilities of such catastrophic events and assessing the effectiveness and tradeoffs of potential risk-reduction measures in light of limited risk-reduction resources and tradeoffs in using them.

Certain types of artificial intelligence (AI) have been proposed as a potentially large factor in GCR. One specific AI type of great concern is artificial superintelligence (ASI), in which the AI has intelligence vastly exceeding humanity's across a broad range of domains (Bostrom 2014). ASI could potentially either solve a great many of society's problems or cause catastrophes such as human extinction, depending on how the ASI is designed (Yudkowsky 2008).

The AIs that exist at the time of this writing are not superintelligent, but ASI could be developed sometime in the future. It is important to consider the long-term possibilities for ASI in order to help avoid ASI catastrophe. With careful analysis, it may be possible to identify indicators that ASI development is going in a dangerous direction, and likewise to identify risk management actions that can make ASI development safer. However, long-term technological forecasting is difficult (Lempert et al. 2003), making ASI risks difficult to characterize and manage. Additional challenges come from the possibility of ASI development going unnoticed (such as in covert development projects) and from weighing the risks posed by ASI against the potential benefits that ASI could bring.

A.M. Barrett (✉) · S.D. Baum
Global Catastrophic Risk Institute, Washington, D.C., USA
e-mail: tony@gcrinstitute.org

V. Callaghan et al. (eds.), *The Technological Singularity*,
The Frontiers Collection, DOI 10.1007/978-3-662-54033-6_6

This paper surveys established methodologies for risk analysis and risk management as they can be applied to ASI risk. ASI risk can be addressed in at least two ways: (1) by building safety mechanisms into the ASI itself, as in ASI "Friendliness" research, and (2) by managing the human process of researching and developing ASI, in order to promote safety practices in ASI research and development (R&D). This paper focuses on the human R&D process because it has similarities to the R&D processes for other emerging technologies. Indeed, the ASI risk analysis ideas presented here are similar to our own work on risks posed by another emerging technology, synthetic biology (Barrett 2014).

The ultimate goal of ASI risk analysis is to help people make better decisions about how to manage ASI risks. Formalized risk methodologies can help people consider more and better information and reduce cognitive biases in their decision making. A deep risk perception literature indicates that people often have grossly inaccurate perceptions of risks (Slovic et al. 1979). One example is in perceptions of "near miss" disasters that are luckily but narrowly avoided. An individual's framing of the near miss as either a "disaster that did not occur" or a "disaster that almost happened" tends to decrease or increase, respectively, their perception of the future risk of such a disaster (Dillon et al. 2014). This, combined with the high stakes of ASI, suggests substantial value in formal ASI risk analysis.

6.2 Key ASI R&D Risk and Decision Issues

For risk analysis, important questions concern the probabilities, timings, and consequences of the invention of key ASI technologies. Regarding the consequences, Yudkowsky (2008), Chalmers (2010) and others argue that ASIs could be so powerful that they will essentially be able to do whatever they choose. Yudkowsky (2008) and others thus argue that technologies for safe ASI are needed before ASI is invented; otherwise, ASI will pursue courses of action that will (perhaps inadvertently) be quite dangerous to humanity. For example, Omohundro (2008) argues that a superintelligent machine with an objective of winning a chess game could end up essentially exterminating humanity because the machine would pursue its objective of not losing its chess game, and would be able to continually acquire humanity's resources in the process of pursuing its objective, regardless of costs to humanity. We refer to this type of scenario as an ASI catastrophe and focus specifically on this for the remainder of the paper.

The risk of ASI catastrophe has the dynamics of a race. Society must develop ASI safety measures before it develops ASI, or else there will be an ASI catastrophe. Estimating ASI catastrophe risk thus requires estimating the probabilities of ASI and ASI safety measures occurring at different times. For ASI invention, a number of technology projection models exist, e.g. The Uncertain Future (Rayhawk et al. 2009a). ASI safety measure models are less well formulated at this point but would be needed for a complete risk analysis.

For risk management, the most important question is: What policies (public or private) should be pursued? A variety of ASI risk reductions policy options have been identified (e.g., Sotala and Yampolskiy 2015; a version of which appears as Part 1 of this volume). At least three sets of policies could be followed, each with its own advantages and disadvantages:

(1) Governments, corporations, and other entities could implement ASI R&D regulations within their jurisdictions, and pursue treaties or trade agreements for external cooperation. Regulations could restrict risky ASI R&D. However, implementation could be costly and could impede benign R&D. It would also be unlikely to be universally agreed and enforced, such that risky research could proceed in unregulated regions or institutions.
(2) Security agencies could covertly target risky ASI projects. Similar covert actions have reportedly been taken against other R&D projects, such as the Stuxnet virus used against Iran's nuclear sector. Such actions can slow down dangerous projects, at least for a while, but they could also spark popular backlash, harden project leaders' desire to continue, and provide dangerous ASI R&D efforts with incentives to avoid detection.
(3) Governments, corporations, foundations, and other entities could fund ASI safety measure development. This could increase the probability of ASI safety measures being available before ASI. However, ASI communities do not have consensus on ASI safety measure concepts or best approaches—more on this below—and some ASI safety measures may still take more time to develop than ASI, in which case ASI catastrophe would still occur.

Here are some potentially important factors:

- Provability of safety measures built into ASI goals ("Friendliness") (Muehlhauser 2013)
- ASI technology development arms race (Shulman and Armstrong 2009)
- Whether and how a government ought to support ASI safety (McGinnis 2010)
- Government intervention blowback risks and other drawbacks (Chalmers 2010, footnote 14)
- Potential for "hard takeoff" versus "soft takeoff" of AI, and their relation to hardware versus software limitations on AI takeoff (Shulman and Sandberg 2010)

6.3 Risk Analysis Methods

6.3.1 Fault Trees

Fault trees represent the ways that events and conditions could combine to lead to a particular outcome. Each node in the tree represents a particular event, such as an attempted use of ASI, or a condition, such as the existence of a new ASI technology. The "top event" node in the tree represents the scenario outcome. Below the top node, the tree branches out with additional nodes. Each layer in the tree represents the combination of events and conditions that could lead to the outcome in the layer directly above it. Nodes are connected by Boolean logic gates, such as OR, AND, and NOT gates, which are an important part of specifying the particular combinations of events and conditions that could result in the outcomes above them in the tree. The tree thus shows a set of possible scenarios, each of whose "fault" it could be for the occurrence of the top event.

Figure 6.1 presents the logic model for a simple ASI catastrophe scenario fault tree. ASI catastrophe occurs if building an ASI is physically possible, if an ASI is built, if the ASI gains "decisive strategic advantage", and if the ASI actions are unsafe. ASI can be built either (1) directly, meaning humans create ASI on their own, or (2) via recursive self-improvement, in which humans build a "seed" AI that builds successively more intelligent AIs until it becomes superintelligent. Decisive strategic advantage means "a level of technological and other advantages sufficient to enable it (the AI) to achieve complete world domination" (Bostrom 2014, p. 78). ASI actions are unsafe if the ASI uses its decisive strategic advantage to cause catastrophe. ASI actions are unsafe if (1) its goals are unsafe, such that it would cause catastrophe if it pursues its goals, and (2) it is not deterred by any humans, other AIs, or anything else, such that it pursues its goals.

Fault tree logic models such as Fig. 6.1 can be extended for quantitative risk analysis using parameters with any real-number value instead of just Boolean logic. Such fault trees can be used to estimate the occurrence rate or probability of the top-event outcome using other rate or probability variables as model inputs. Essentially, quantitative models are created by adding quantitative values for model parameters (often either rate or probability variables) in the fault tree logic models. Each node represents a variable with an associated rate (e.g. a rate of origination of entities that would attempt to create ASI if they obtain sufficient resources) or probability (e.g. a probability that a new ASI technology would be available to

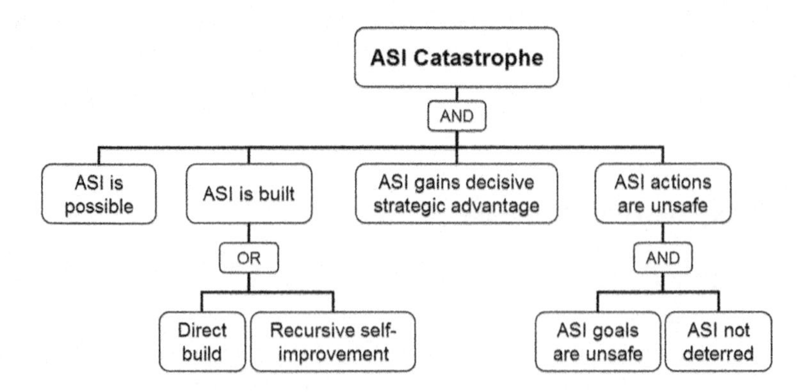

Fig. 6.1 Simple ASI catastrophe fault tree logic model

entities at a particular point in time). For mathematical details and an example of how rate and probability variables can be combined in a risk model fault tree, see Barrett et al. (2013). Quantitative models provide better comparison of risk magnitudes than logic models, but building quantitative models is harder, as it requires more data and/or assumptions regarding parameter values.

6.3.2 Event Trees

Event trees represent the possible outcomes of an event, and the probabilities of arriving at each of those outcomes. Event trees can be used to represent at least two kinds of important ASI catastrophe risk factors. The first and simplest application is to represent the probabilities of either the current (unknown) state, or the future state, of a particular condition. For example, an important ASI catastrophe risk factor is whether it is fundamentally possible to invent an ASI. Although some individuals may have opinions about whether the proposition is true, it is unknown at this point in time whether it is true. As shown in Fig. 6.2, the probability of the proposition being true can be represented by the parameter p_T, and the probability of the proposition being false can be represented by the quantity $1 - p_T$.

A second application of event trees is for technology development modeling, working forwards from the current state of the world. ASI technology development models can provide estimates of the probabilities of ASI development conditions as a function of time and potentially other variables (e.g. the financial resources of actors pursuing ASI).

Important conditions include (1) whether a technology has been invented or made available and (2) how affordable the technology is. These two conditions are for the forecasting of ASI developed via "grind" or "insight" (Armstrong and Sotala 2012). Grind involves applying established techniques repeatedly for gradual progress. Examples of grind include gradual progress in semiconductor manufacturing permitting faster hardware per dollar (as in Moore's law) and gradual progress in brain imaging permitting more detailed artificial brain emulations (as in whole brain emulation; see Sandberg and Bostrom 2008). Insight involves intellectual breakthroughs bringing fast, transformative progress. An example of insight is if ASI requires advances in algorithms that would not need advanced hardware to run on. Figure 6.3 shows an event tree for when an ASI "insight" breakthrough could occur, with each year between 2015 and 2100 as a possibility, and with a probability parameter for each year.

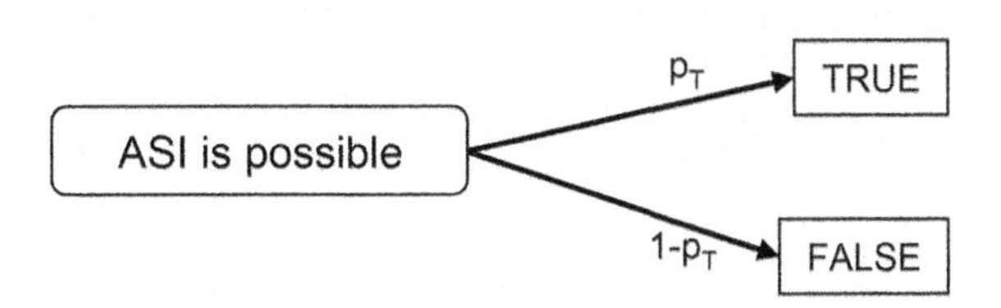

Fig. 6.2 Event tree of whether ASI is fundamentally possible

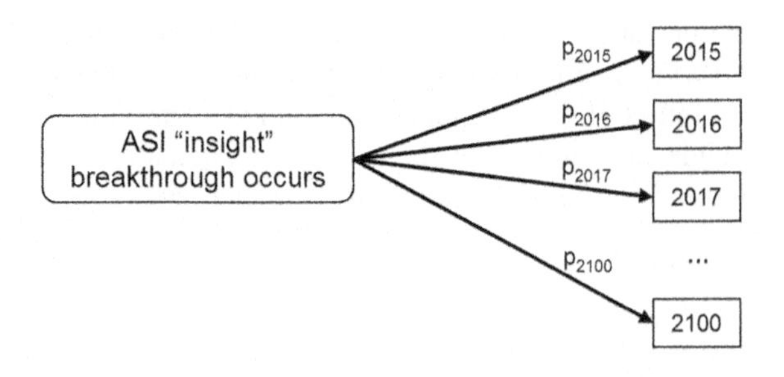

Fig. 6.3 Event tree of year when "insight" breakthrough occurs

6.3.3 Estimating Parameters for Fault Trees and Event Trees

Using fault trees and event trees for quantitative risk analysis requires estimating parameter values for all model components. Parameter estimation is straightforward when parameter values are known. However, parameter values are often uncertain, in which case techniques are needed for characterizing the uncertainty, typically in terms of probability distributions.

Many risk analyses form parameter probability distributions based on some combination of empirical data and expert judgment. Depending on the type of variable that the parameter represents, the parameter can often be approximated by one of several well-known probability distributions. For example, parameters representing rates or frequencies (holding any positive value) sometimes assume a Poisson process (e.g. for randomly-timed ASI creation events), in which case gamma distributions can be used; for parameters representing binomial processes (which have two possible outcomes: true/false, yes/no, etc.), beta distributions can be used (Bolstad 2007). For parameters representing multinomial processes (with more than two possible outcomes), Dirichlet distributions can be used.

For technological forecasting, linear regressions are often used to extrapolate trends, especially over the relatively short term, e.g. a period of one year (Millett and Honton 1991, pp. 13–14). Note that linear regressions can be used for transformations of raw data. The most notable example of this is the exponential progression seen in Moore's law: a plot of log (performance) versus cost gives a fairly straight line (Stokes 2008). Lognormal distributions have been used for estimating invention date. For example, Rayhawk et al. (2009b) postulate a lognormal distribution for the number of years until there is enough brain imaging technology to build neuromorphic artificial general intelligence. Fallenstein (2013) suggests a Pareto distribution for ASI arrival date. Modis (2012) and Kurzweil (2012) debate the use of linear regressions and logistic curves. Sandberg (2010) surveys a number of distributions and models that have been suggested in various ASI technology development models.

The probability distributions, whether based on empirical data or expert judgment, can be interpreted in Bayesian terms. Expert judgment can serve as a prior distribution, to be updated as empirical data is observed. In the absence of any expert judgment, uniform distributions can be used to represent ignorance about a parameter value, though care must be taken here because uniform distributions are sensitive to the choice of model structure (Pratt et al. 1995, pp. 236–237).

Finally, the passage of time can lead to further updating. In some cases, if time has passed and no new indicators have been observed, that can be counted as evidence for or against particular hypotheses, which itself should count as evidence for use in Bayesian updating.

6.3.4 Elicitation of Expert Judgment

Because ASI is an unprecedented technology, and because it may currently be at an early stage of development, there will be no empirical data for large portions of the parameters of any fault tree or event tree model of ASI R&D. To estimate these parameters with something more than just a uniform distribution, expert judgment is needed.

Expert judgment has significant limitations across all domains of expertise, including for AI predictions (Armstrong and Sotala 2012; Armstrong et al. 2014). However, expert elicitation best practices can help overcome the limitations. For example, actual experts should exist for the questions asked, and models or elicitation questions should be structured to take advantage of experts' knowledge (Morgan and Henrion 1990, pp. 128–137; Meyer and Booker 1991, pp. 24–26). Unfortunately, many expert judgments about AI have not used the best practices (Armstrong and Sotala 2012, sec. 4.1; Armstrong et al. 2014). For example, many AI predictions consist only of point estimates of when specific AI milestones will be accomplished; a more complete characterization of the uncertainty about AI milestone timing suggests using probability distributions instead of point estimates (Baum et al. 2011).

As argued by Armstrong and Sotala (2012), AI technological progress forecasts appear to have often been substantially wrong, even when made by "experts". However, ASI experts may not be the most knowledgeable individuals about technological forecasting, and vice versa. Thus better models for ASI technological forecasting may be constructed using expert elicitations of combinations of threat-domain experts and tech-forecasting experts as follows:

- Use ASI experts to inform model structure (e.g. the nature and number of major technological development steps necessary for ASI)
- Use technological forecasting experts to inform model parameters (e.g. quantities of time and resources typically required for major technological developments)

In addition, where possible, forecasting models should be structured to include intermediate-step claims for empirical testing and updating. That should help prevent excessive reliance upon experts without opportunities to check their forecasts.

6.3.5 Aggregation of Data Sources

In risk modeling, mathematical methods are often used to aggregate empirical data and expert judgment in order to arrive at a model that represents the best information available. However, such methods should not be used unthinkingly—there are multiple ways of conducting mathematical aggregation, and sometimes it is better to leave data disaggregated (Keith 1996). For example, differences in experts' views may result from important differences in their fundamental assumptions; aggregating their views hides these differences, to the detriment of risk analysis and management. Morgan and Henrion (1990) advise using model analysis methods to identify the most important input factors in a model, and then to explore and communicate the key points about the model's dependence on values of key input factors, rather than simply trying to merge all input values together using aggregation methods.

In cases where aggregation of judgments from an expert elicitation process is appropriate, e.g. in combining differing opinions among experts with roughly similar fundamental assumptions, weighted aggregation is often performed using Bayesian statistics. We focus on weighting for beta distributions in the following. Basic Bayesian aggregation for beta distributions (for a binomial process) is described in Meyer and Booker (1991, pp. 331–335). The basic method assumes that each of several experts provides a judgment about beta distribution parameter values y and n, where y is the number of successes in n trials. Thus each expert i provides values y_i and n_i. Then aggregation of the judgments provides aggregated values y' and n', where $y' = \sum_i y_i$ and $n' = \sum_i n_i$. The basic approach also could be fairly simple to use for adjustable weighting of information from various sources. Weighting is used sometimes in processing expert elicitation judgments when not all experts are regarded as being equally credible. When one expert is viewed as being more credible than another, their views are given higher weighting. It is simple to extend the above-mentioned aggregation process to account for weights, by giving each expert's judgment a weight w_i. Then weighted aggregation of the judgments uses the formulas $y' = \sum_i w_i y_i$ and $n' = \sum_i w_i n_i$.

Expert judgment performance-based methods are sometimes used in deciding how much weight to give to judgments made by different experts in an expert elicitation, as in the expert judgment performance calibration methods of Cooke (1991) where the elicitor assesses the performance of experts based on how well they respond to questions with answers known to the elicitor (O'Hagan et al. 2006, pp. 184–185).

6.4 Risk Management Decision Analysis Methods

The point of risk analysis is, in general, not the analysis itself, but its potential to inform risk management decisions. For ASI catastrophe, there are many risk management options, which are available for a variety of decision makers (Sotala and Yampolskiy 2015). Relevant decision makers for ASI R&D include governments and non-governmental organizations active in ASI oversight, as well anyone sponsoring, conducting, or otherwise supporting ASI R&D. All of these individuals and groups have roles to play in addressing ASI risk. Ideally an ASI decision analysis would inform all of these many decisions, though in practice analysts must focus on only some decisions.

In simplest terms, risk management decisions are evaluated according to two factors: the options and the objectives. The options constitute the set of all possible risk management actions, including the act of doing nothing. In practice, there is often an infinity of possible actions. To make analysis tractable, one must select a finite portion of these options. When each option is evaluated by hand, a small number of options must be chosen, with each ideally representing some important class of options. For ASI R&D, important classes of options include abstaining from developing ASI ("relinquishment"; Joy 2000) and conducting ASI safety measures research.

The objectives for risk management decisions are the underlying goals or purposes that the decision makers seek to accomplish. The objectives can be expressed in terms of an objective function, optimization criterion, utility function, social welfare function, ethical framework, or similar analytic paradigms. These paradigms have implicit commensurability between all items being valued (e.g. lives saved vs. dollars spent), which allows for a relatively simple equation for the expected value of a variety of activities types and their consequences (Clemen and Reilly 2001, p. 512). Other decision analysis research uses multi-criterion objective functions, seeking to identify options that perform well across a range of objectives (Keeney and Raiffa 1976).

ASI decision analysis is complicated by the prospect of including the ASI itself as an intrinsically valuable objective, i.e. an objective that is worth pursuing for its own sake, without reference to other objectives. The philosophical basis of many objective functions is the view that it is intrinsically valuable to satisfy the preferences or improve the subjective experience of sentient beings (e.g., Broome 1991)—and all sentient beings, not just humans (Ng 1995). If an ASI is sentient, then its preferences or experiences arguably ought to be counted too. The question of whether an ASI can be sentient is very difficult to answer, touching on deep questions in the philosophy of mind (Chalmers 2010). If an ASI would be sentient, then its preferences or experiences potentially could be an important factor in a decision analysis.

Another complication for ASI decision analysis is the extremely high stakes. An AI could either solve a great many of society's problems or cause human extinction (Yudkowsky 2008). Either of these outcomes could dominate a typical decision analysis (Beckstead 2013). Published estimates of the value of preventing human extinction vary wildly, from $600 trillion (Posner 2004) to infinity (Weitzman 2009; Baum 2010). The value of solving society's problems could be at least as large.

One way to sidestep these complications is by using cost-effectiveness analysis (CEA). CEA seeks options that achieve some fixed objective for the lowest possible cost. For ASI risk management, a fixed objective could be avoiding ASI catastrophe. This objective can then be pursued regardless of how valuable it is. Likewise, regardless of how valuable is, society only has finite, limited resources available for reducing this risk. CEA can help identify how to allocate the resources to minimize ASI catastrophe risk.

To illustrate CEA of ASI risk management, consider the two options of relinquishment and Friendliness research. Major world governments may be able to pursue total relinquishment of ASI development. This might greatly reduce ASI catastrophe risks, but it could also be expensive to enforce and could have a large opportunity cost due to foregone benefits of ASI and related types of narrow AI. These costs reduce the merits of total relinquishment and could make governments less likely to pursue it. In comparison, ASI safety measures research may also reduce ASI catastrophe risks, and at much smaller cost than total relinquishment. Depending on the details, ASI safety measures research could be more cost-effective than total relinquishment.

6.5 Evaluating Opportunities for Future Research

Decision analysis can be of further help for guiding future research. The basic idea is that it is often helpful to design a research project in consideration of the decisions it can inform. From a decision perspective, research is of higher value when it better improves the performance of decisions by reducing decision uncertainty. High value research thus occurs when it brings decision makers new information that is relevant to the decisions at hand. Valuations of research can in turn inform decisions on the allocation of resources to various lines of research. Some prior research has considered the value of ASI risk research (Salamon 2009; Yudkowsky 2013, pp. 82–84).

The decision analysis concept Expected Value of Perfect Information (EVPI; Clemen and Reilly 2001, p. 512), can provide a formal quantitative approach to assessing the value of ASI risk research. EVPI is the difference between the expected value of a decision with perfect information versus with only currently available information. In the context of a decision analytic model, such as a fault tree or an event tree, the expected value of information is based on the extent to which information reduces the uncertainty about the value of a particular parameter in the model. Perfect information about a parameter eliminates that uncertainty.

In general, EVPI calculations are used to set an upper limit to how much should be spent on reducing uncertainty—research cannot produce better-than-perfect information. On their own, EVPI calculations cannot predict how valuable specific research will be in reducing uncertainty. However, even imperfect information can have great value in reducing decision model parameter uncertainties by some amount. Straightforward extensions of approaches to EVPI calculations can provide methods to assess the Expected Value of Imperfect Information (EVII; Clemen and Reilly 2001) and Expected Value of Including Uncertainty (EVIU; Morgan and Henrion 1990).

As with decision analysis in general, the expected value of information is typically evaluated using utility functions or functionally similar metrics. This introduces the same complications of evaluating the high stakes of ASI decisions. Barrett and Baum (2014) provide an approach to estimating value of information based on cost effectiveness that avoids these complications. This approach can be helpful for evaluating AI risk research.

6.6 Concluding Thoughts

We believe there is significant value in working towards a single integrated model that can represent the most important policy-relevant ASI catastrophe risk factors, and incorporate the best information available about those factors, all at a tractable level of detail. However, that model will not serve all purposes for all stakeholders. It is not tractable or useful to try to build a single model that tries to answer all potential questions, nor that tries to model all potential issues at an extremely high level of detail.

Even if an analysis cannot produce rigorous quantitative answers to decision questions, it can still be worth conducting. The risk analysis literature suggests that often, the most useful outcomes of a probabilistic risk analysis modeling effort are often not the model's outputs themselves (which may or may not be surprising to experts), but the new insights and improved communication regarding risks and risk management strategies that result from the structured thinking and multidisciplinary discussions needed for the analysis (Kumamoto and Henley 1996, pp. 132–136).

An important qualification that is important to recognize is that some aspects of ASI research could actually increase risks. The research thus should appropriately protect sensitive information, while providing description of methods sufficient to allow other researchers to examine and employ them. In general, we suggest protecting from general publication information that is non-obvious, not easily available from other sources, and that would be useful to actors with ill intent or with significant capacity for inadvertently causing harm. Similar rules of behavior for researchers are typically used in security-sensitive government work, and are being increasingly used in academic research.

In summary, risk analysis and decision analysis methods offer time-tested approaches to structuring assessment of risk issues to allow well-informed, transparent risk management decisions. This holds for emerging technology risks like ASI just as it does with other types of risk. The methods do not necessarily offer a "right" way of proceeding, and they are not purely a science; there is an art involved, with best practices suggested by empirical studies. However, risk and decision analyses can serve several important purposes. One purpose is to help clarify key technical issues (both what is known and what could be learned with further research) in context of important decisions. Another purpose is to help separate technical issues from matters of value and policy debates. Both of these could be of great value for ASI R&D risks, given the range of stakeholders and potentially very high stakes involved.

Acknowledgements Thanks to the editors for comments on the chapter manuscript, and Stuart Armstrong, Luke Muehlhauser, Miles Brundage, Roman Yampolskiy, and others for comments on related research. Any opinions, findings and conclusions or recommendations in this document are those of the authors and do not necessarily reflect views of the Global Catastrophic Risk Institute, nor of others.

References

Armstrong, S. and K. Sotala (2012). How We're Predicting AI—or Failing To. Beyond AI: Artificial Dreams. Pilsen, Czech Republic, University of West Bohemia: 52–75.

Armstrong, S., K. Sotala and S. S. Ó hÉigeartaigh (2014). "The errors, insights and lessons of famous AI predictions – and what they mean for the future." *Journal of Experimental & Theoretical Artificial Intelligence*.

Barrett, A. M. (2014). "Analyzing Current and Future Catastrophic Risks from Emerging-Threat Technologies." Retrieved May 5, 2014, from http://research.create.usc.edu/cgi/viewcontent.cgi?article=1062&context=current_synopses.

Barrett, A. M. and S. D. Baum (2014). Value of GCR Information: Cost Effective Reduction of Global Catastrophic Risks (GCRs). Advances in Decision Analysis. Washington, D.C.

Barrett, A. M., S. D. Baum and K. R. Hostetler (2013). "Analyzing and Reducing the Risks of Inadvertent Nuclear War Between the United States and Russia." *Science & Global Security* 21 (2): 106–133.

Baum, S. D. (2010). "Is humanity doomed? Insights from astrobiology." *Sustainability* 2(2): 591–603.

Baum, S. D., B. Goertzel and T. G. Goertzel (2011). "How long until human-level AI? Results from an expert assessment." *Technological Forecasting & Social Change* 78(1): 185–195.

Beckstead, N. (2013). On The Overwhelming Importance Of Shaping The Far Future. Department of Philosophy. New Brunswick, New Jersey, Rutgers University.

Bolstad, W. M. (2007). Introduction to Bayesian Statistics. Second ed. Hoboken, New Jersey, John Wiley & Sons.

Bostrom, N. (2002). "Existential Risks: Analyzing Human Extinction Scenarios and Related Hazards." *Journal of Evolution and Technology* 9(1).

Bostrom, N. (2014). Superintelligence: Paths, dangers, strategies. Oxford, United Kingdom, Oxford University Press.

Broome, J. (1991). "Utility." *Economics and Philosophy* 7(1–12).

Chalmers, D. (2010). "The Singularity: A Philosophical Analysis." *Journal of Consciousness Studies* 17: 7–65.

Clemen, R. T. and T. Reilly (2001). Making Hard Decisions. 2nd ed. Pacific Grove, California, Duxbury.

Cooke, R. (1991). Experts in Uncertainty: Opinion and Subjective Probability in Science. New York, Oxford University Press.

Dillon, R. L., C. H. Tinsley and W. J. Burns (2014). "Near-Misses and Future Disaster Preparedness." *Risk Analysis*.

Fallenstein, B. (2013). Predicting AGI: What can we say when we know so little? Berkeley, California, Machine Intelligence Research Institute.

Joy, B. (2000). "Why the future doesn't need us." Wired Retrieved 9 October 2011, from http://www.wired.com/wired/archive/8.04/joy.html.

Keeney, R. and H. Raiffa (1976). Decisions with Multiple Objectives: Preferences and Value Trade-Offs. New York, John Wiley & Sons, Inc.

Keith, D. W. (1996). "When is it appropriate to combine expert judgments?" *Climatic Change* 33 (139–144).

Kumamoto, H. and E. J. Henley (1996). Probabilistic Risk Assessment and Management for Engineers and Scientists. 2nd edition ed. New York, IEEE Press.

Kurzweil, R. (2012). On Modis' "Why the Singularity Cannot Happen". Singularity Hypotheses: A Scientific and Philosophical Assessment. A. H. Eden, J. H. Moor, J. H. Soraker and E. Steinhart. New York, Springer: 343–348.

Lempert, R. J., S. W. Popper and S. C. Bankes (2003). Shaping the Next One Hundred Years: New Methods for Quantitative, Long-Term Policy Analysis. Santa Monica, California, RAND.

Maher Jr., T. M. and S. D. Baum (2013). "Adaptation to and recovery from global catastrophe." *Sustainability* 5(4): 1461–1479.

McGinnis, J. O. (2010). "Accelerating AI." *Northwestern University Law Review* 104: 366–381.

Meyer, M. A. and J. M. Booker (1991). Eliciting and analyzing expert judgment: a practical guide. London, Academic Press Limited.

Millett, S. M. and E. J. Honton (1991). A manager's guide to technology forecasts and strategy analysis methods. Columbus, Ohio, Battelle Press.

Modis, T. (2012). Why the Singularity Cannot Happen. Singularity Hypotheses: A Scientific and Philosophical Assessment. A. H. Eden, J. H. Moor, J. H. Soraker and E. Steinhart. New York, Springer: 311–340.

Morgan, M. G. and M. Henrion (1990). Uncertainty: A guide to dealing with uncertainty in quantitative risk and policy analysis. Cambridge, Cambridge University Press.

Muehlhauser, L. (2013). "Mathematical Proofs Improve But Don't Guarantee Security, Safety, and Friendliness." Retrieved May 10, 2014, from http://intelligence.org/2013/10/03/proofs/.

Ng, Y.-K. (1991). "Should we be very cautious or extremely cautious on measures that may involve our destruction?" *Social Choice and Welfare* 8: 79–88.

Ng, Y.-K. (1995). "Towards welfare biology: evolutionary economics of animal consciousness and suffering." *Biology and Philosophy* 10: 255–285.

O'Hagan, A., C. E. Buck, A. Daneshkhah, J. R. Eiser, P. H. Garthwaite, D. J. Jenkinson, J. E. Oakley and T. Rakow (2006). Uncertain Judgements: Eliciting Expert Probabilities. Chichester, West Sussex, England, John Wiley & Sons.

Omohundro, S. (2008). The Basic AI Drives. Proceedings of the First AGI Conference, Frontiers in Artificial Intelligence and Applications. P. Wang, B. Goertzel and S. Franklin, IOS Press. 171.

Posner, R. A. (2004). Catastrophe: Risk and Response. New York, Oxford University Press.

Pratt, J. W., H. Raiffa and R. Schlaifer (1995). Introduction to Statistical Decision Theory. Cambridge, Massachusetts, MIT Press.

Rayhawk, S., A. Salamon, T. McCabe, M. Anissimov and R. Nelson (2009a). Changing the frame of AI futurism: From storytelling to heavy-tailed, high-dimensional probability distributions. 7th European Conference on Computing and Philosophy (ECAP). Bellaterra, Spain.

Rayhawk, S., A. Salamon, T. McCabe, M. Anissimov and R. Nelson (2009b). "The Uncertain Future." Retrieved 23 October 2011, from http://www.theuncertainfuture.com/.

Sagan, C. (1983). "Nuclear war and climatic catastrophe: Some policy implications." *Foreign Affairs* 62(2): 257–292.

Salamon, A. (2009). How Much it Matters to Know What Matters: A Back of the Envelope Calculation. Singularity Summit 2009. New York.

Sandberg, A. (2010). An overview of models of technological singularity. Workshop on Roadmaps to AGI and the future of AGI (at AGI10 conference). Lugano, Switzerland.

Sandberg, A. and N. Bostrom (2008). Whole Brain Emulation: A Roadmap, Future of Humanity Institute, Oxford University. Technical Report #2008-3.

Shulman, C. and S. Armstrong (2009). Arms control and intelligence explosions. 7th European Conference on Computing and Philosophy (ECAP). Bellatera, Spain.

Shulman, C. and A. Sandberg (2010). Implications of a Software-Limited Singularity. ECAP10: VIII european conference on computing and philosophy. K. Mainzer. Munich, Verlag Dr. Hut.

Slovic, P., B. Fischhoff and S. Lichtenstein (1979). "Rating the Risks." *Environment* 21(3): 14–20, 36-39.

Sotala, K. and R. V. Yampolskiy (2015). "Responses to Catastrophic AGI Risk: A Survey." *Physica Scripta* 90(1).

Stokes, J. (2008). "Understanding Moore's Law." Retrieved January 25, 2014, from http://arstechnica.com/gadgets/2008/09/moore/.

Weitzman, M. L. (2009). "On modeling and interpreting the economics of catastrophic climate change." *Review of Economics and Statistics* 91(1): 1–19.

Yudkowsky, E. (2008). Artificial intelligence as a positive and negative factor in global risk. Global Catastrophic Risks. N. Bostrom and M. M. Cirkovic. Oxford, Oxford University Press: 308–345.

Yudkowsky, E. (2013). Intelligence Explosion Microeconomics. Berkeley, California, Machine Intelligence Research Institute.

Chapter 7
Diminishing Returns and Recursive Self Improving Artificial Intelligence

Andrew Majot and Roman Yampolskiy

7.1 Introduction

The concept of a recursively self-improving intelligence has been around for quite a while (Good 1965). At its core is the belief that the creation of a human or superhuman intelligence level is far over the horizon of current science, but a more modest and simple intelligence could be possible within a reasonable amount of time. This reasonable timeframe is unfortunately one of the best examples of a moving target that computer science has ever produced. Initial estimates in the 1960s were that this kind of intelligence, or for that matter any kind of intelligence, might be possible by 1980. When this proved not to be the case the estimate was moved again and again. Some of the current estimates have taken lesson from previous attempts at forecasting AI developments and put the estimate at any time between 20 and 100 years (Sotala and Yampolskiy 2013). The main reasons for this constantly shifting estimate is our continued floundering in the creation of any remotely workable models for cognition. We can describe overall cognitive functioning in some detail, but have no real concept behind things as basic as what constitutes an artificial thought or how to code a machine with common sense. These basic roadblocks to intelligence are open problems whose solutions we don't even have an inkling of yet.

The idea of a "Seed AI" capable of recursive self-improvement was created to get around these bottlenecks. It seems inherently more difficult to create a super-intelligent AI from the start than it is to let a less intelligent one figure out

A. Majot (✉) · R. Yampolskiy
Department of Computer Engineering and Computer Science,
University of Louisville, Louisville, USA
e-mail: amajot@gmail.com

R. Yampolskiy
e-mail: roman.yampolskiy@louisville.edu

V. Callaghan et al. (eds.), *The Technological Singularity*,
The Frontiers Collection, DOI 10.1007/978-3-662-54033-6_7

how to improve itself. The human intellect is currently barely able to begin comprehending itself, and the thought of it being able to generate something far more intelligent than itself is a bit far fetched. So instead of solving every single problem related to creating super-intelligence why don't we have the AI itself do it? That is the main proposal behind Seed AI. Humans would only need to produce one AI capable of a modicum of cognition, programmed with the drive to improve itself, and capable of some degree of original thought and imagination. Over time this AI would examine its own source code, picking up various algorithms it has and analyzing them to look for weaknesses. Once identified, these weaknesses would be improved upon and the new algorithm tested, with the AI modeling itself and its new potential. If the algorithm works and produces improvement, the AI would edit its source code to include the upgraded component. This would be done ad infinitum until a limit was reached, or the AI chooses to stop improving itself and instead embark on a new quest. We will go over the details of how this recursive improvement would work later in the chapter, and show that some of the methods used for this process have some possible limitations.

Admittedly, this whole concept is consigned to the semi-distant future, so why do we need to worry ourselves over it now? There are many people who would argue that our time and resources are much better spent on other problems (McDermott 2012). Others think that once this type of intelligence is created, it would have the ability to solve all of the most pressing and critical to survival problems that we currently face (Chalmers 2010; Yudkowsky 2007). Such an intelligence would have a wholly unique view on any problems we face. It would probably not think like a human, have the biases inherent in any human culture, or have the same motivations as a human. This alone makes an artificial intelligence an amazing asset capable of independent analysis and thought. Even if it doesn't solve all of our problems overnight, such a being should be valued as an ultimate way to gain inhuman perspectives on human problems. Who knows what kind of insights a circuit-based life form can provide to us carbon-based life forms who struggle daily just to play nicely with one another.

The ideal and most helpful example of an artificial intelligence would be as general as possible in order to mimic the capacities of the human brain. Therefore this AI falls under the Artificial General Intelligence (AGI) category (Yampolskiy and Fox 2012; Goertzel and Pitt 2012). One could make any type of intelligence you wished, chess playing, weather forecasting, disease curing, bomb designing, genocide planning, or election rigging. But the most useful kind of intelligence would likely be of the AGI category since it would be able to draw upon any variant of knowledge or deduction to solve problems. This general-purpose thinking AI probably wouldn't have many of the shortcomings that human thinkers have. They could remain focused on their goal, have no need of sleep, and depending on how they are set up, they would have access to the bulk of human knowledge if allowed access to the internet. An internet hookup would certainly be a risky move however. It may be best to ensure a somewhat guided approach to the AI's knowledge acquisition, as there have been recent examples where unintended consequences result from bulk information ingestion. One of the more recent, and humorous examples can be found from IBM's Watson super computer. In order to get Watson more comfortable with the nuances in human language, its programmers crawled the site urbandictionary.com. This had the unintended side effect of making Watson curse like a sailor because it could not tell the difference between polite conversation and crude jokes (Lewis 2013).

The main fear with a recursively improving AGI is that its strengths could exploit our own weaknesses. That is to say that they would quickly outpace us in brainpower while also not having a moral code to inhibit their actions and motivations. These actions and motivations might not necessarily have humanity's best interests in mind. We fear that the AGI's recursive growth could be exponential and within months of being turned on it could have control of the planet (Yampolskiy 2015b; Sotala and Yampolskiy 2013; Omohundro 2008; Turney 1991; Goertzel and Pitt 2012; Armstrong et al. 2012). Omohundro specifically details what could drive an AI towards self-improvement and why it would want to do so. But what if we are looking for solutions to a problem that wouldn't actually exist? Is it really possible for an AI to improve upon itself so drastically that our intelligence would be to it like ants are to us?

To summarize: we show that this scenario is not likely, at least under current mathematical assumptions coupled with a dash of common sense.

- Yes, the AI could improve upon itself, creating novel algorithms and raising its Machine Intelligence Quotient (MIQ) quite a bit, but it would still be bound by inherent limits of algorithm complexity.
- The law of diminishing marginal returns would probably take effect, causing greater and greater computational resource usage for smaller and smaller performance gains.
- Any AI with a goal will need to work towards this goal, limiting the time available for self-improvement since after a certain point spending time on improvement would become prohibitive to completing its goal.
- The logistical chain for creating computers from initial mineral mining and refining to design and construction naturally allows for opportunities to limit an AI bent on runaway hardware improvement.

7.2 Self-improvement

The concept of recursive self-improvement was first mentioned in 1965 by John Irving Good. He postulated that as soon as a reasonably intelligent machine was created, it would go about improving upon itself. It would then inhabit this new improvement, and use the increase in efficiency and ability to create an even better version of itself. This would continually take place until the machine's intelligence would be completely inconceivable to us lowly humans. The machine would be able to solve nearly any current problem bothering mankind, and so would be the "last invention that man need ever make" (Good 1965). What might be the path for this AI to improve upon itself? In order to understand how an AI like this may be developed it might be a good idea to take a look at some potential methods for its initial creation, and how it could go about improving itself.

7.2.1 Evolutionary Algorithms

One example of self-improving algorithms is found in evolutionary algorithms. In these algorithms the root source code stays the same, but any parameters the code is given can be changed, thus improving performance. These algorithms mimic the natural process of evolution within code by generating a population of algorithms or problem solutions, mutating and "breeding" them, and only allowing the fittest solutions to pass on to the next generation. This idea has been around for a while, and has successfully solved many difficult problems that may have taken much longer to solve otherwise (Yampolskiy 2015a; Eiben and Smith 2003).

The components of most evolutionary algorithms are as follows:

- Creation of Initial Population of algorithms or solutions
- Combinatorial/breeding function
- Mutation function
- Selection (Fitness Test)

While certainly not the only means of producing new, better iterations of intelligence, an evolutionary implementation would seem to make the most sense for an AI to better itself. The AI would initially create several candidate improvement algorithms and combine/mutate them according to need. Piecing together what we have learned so far, the determination of the MIQ score for a given self-improving AI would probably be the fitness test in this evolutionary setup. But in order to fully test the MIQ score for an AI, it would need to be able to simulate the various algorithms in its candidate population. This would require extensive resources to be able to run a full or even a limited simulation and, unless the AI has a plethora of resources, would need to be carefully scheduled with existing goals and duties so that they are not adversely affected. Once the simulation is loaded a MIQ test can take place, and the rendered score could be used for the evolutionary selection and fitness test. Only the top results would be allowed to move to the next generation, where they would be cross-bred and further mutated for evolutionary simulation. Perhaps if the simulations provided guaranteed improvement the new algorithm(s) would be loaded as the default cognitive algorithms for the AI. This would allow the next generation of candidates to be created faster, as the AI would run smarter and/or more efficiently. Another key part of this process would be the mutation algorithm. If the AI could guide mutation away from obviously dead ends and prune the possible evolutionary tree it would speed up generational improvement. Theoretically, over time this whole process would create intelligences much more capable than the original.

7.2.2 *Learning Algorithms*

Other self-improving algorithms could be powered by a learning based algorithm. There has already been some research for self-improving and self-optimizing algorithms using this approach (Ailon et al. 2011; Lin 1992). In basic machine learning algorithms the agent is given a base learning algorithm and a number of training data sets. Along with the data sets, expected processed results are also provided so that the agent can learn to associate certain types of input with certain outputs. These training data sets allow the agent to fine-tune itself to a point that it can handle new data sets and reliably provide accurate results. The underlying algorithms behind this learning process can be artificial neural nets, association rule learning, inductive logic programming, support vector machines, bayesian networks, representation learning, and several other methods (Mitchell 1997). Each of these methods has their pros and cons, and would need to be weighed accordingly. Ideally the best methods would be unsupervised learning since handholding an AI may be slow going for initial improvement.

It would seem reasonable for an AI to have the ability to improve by learning for each of its algorithms responsible for cognition. If each function has the ability to learn and improve its own performance, then all of their combined improvements would theoretically increase the MIQ of the AI. But that may just increase performance, it might not provide the imaginative leaps needed for sustained improvement over time. To compensate for this a function separate from the base AI intellect may be called for. This function could analyze cognitive algorithms, their performance, and their effectiveness and attempt to generate new versions. These new versions could then be simulated and the results benchmarked, with the function learning from its mistakes and successes to create better and better versions of cognitive algorithms. Eventually this approach may lead to the leaps and breakthroughs needed to avoid premature, stagnant dead ends.

7.3 Limits of Recursively Improving Intelligent Algorithms

Most of the fears behind constantly improving AIs revolve around the assumption that they will get super intelligent rather quickly. This also assumes that the algorithms employed will be so complex and daunting that humans would not be able to understand their structure or purpose, creating a sort of black box AI. This takes for granted a couple of important things in order for this scenario to be plausible. First, humanity is not the ultimate form of intelligence, and there exists possible intelligences that are better than ours. Here the authors do not disagree with this notion, every time you turn on the television and watch the news there seems ample evidence that human intelligence can be improved upon. The second assumption is that intelligence can be infinitely improved upon, or improved upon in such a way as to have a runaway improvement curve that humanity would be hard pressed to follow. Here we do have an issue with the assumption, which we will detail along with our reasoning. First though we need to think about what kinds of improvements an AI can look for to boost its MIQ. Broadly speaking these improvements can fall under two categories, software and hardware.

7.3.1 Software Improvements

According to a number of previously referenced definitions, intelligence can be described and measured as several discrete components acting together to form an intelligent being. Examples include components that handle sensory input like visual or auditory cues, ones that can perform counting or mathematical functions, and others that can make decisions based on input, previous experiences and available data. In nature these separate functions can be described in the various regions of the human and animal brain. Taken individually each of these regions can perform specific functions that contribute to overall intellect. Taken together they create something that is much larger and more functional than just the simple sum of its parts. The regions and discrete functions within the brain interact in complex and currently not well understood ways. This is evidenced by many of the most successful intelligence tests that query the individual test subject in a variety of separate categories. Spatial and verbal reasoning, logical deduction, and mathematical prowess are all measured in standard IQ tests. Each of these has a region of the brain associated with it (Jung and Haier 2007). It would make sense for an AI to behave in the same manner, with discrete algorithms for individual functions that, when combined, form intelligence much greater than the simple sum of its parts.

Since the only current examples of a working intelligence to be found are ourselves, and to a lesser extent our animal relatives, it makes sense to start the quest for AI by mimicking the intellectual structures found in intelligent creatures. Some have proposed doing just that (Yudkowsky 2007). Our seed AI would be loaded with a full complement of individual algorithms whose full source code is available to itself and it would then be tasked with improving itself. Our initial attempts at creating this base level AI would tie these basic algorithms together into a rough semblance of cooperation. This would be just enough intelligence to provide a functional cohesion capable of limited intellect. Then the AI would churn through computational cycles modeling improvements for each of its algorithms, along with improvements on how each of these algorithms would interact. After an indeterminate amount of time it would come up with a better version of itself and then install that new software and run using that upgrade, thus improving a sector of its intelligence by a fixed amount. This would be repeated over and over, until the AI reaches and surpasses our level of intelligence.

But here we find ourselves with a problem. Using this collection of algorithms approach we would eventually run out of improvements to find for each individual algorithm. Algorithms would be improved to such a point that they would become provably the best at what they do (Legg 2008; McDermott 2012). From a software perspective there would only be a couple of options. Hunt for better combinations/interactions of algorithms, or create novel, additional algorithms that would expand functionality. The combination of both does have some interesting possibilities with potentially near-infinite permutations. But what would this actually look like? The AI would gain talents and abilities, but would it really be any more intelligent? For example, let us say that it were to formulate and then improve to maximum efficiency an algorithm capable of modeling net energy gain nuclear fusion in a Tokomak reactor. This would allow the AI to probe new insights into nuclear fusion, potentially solving a very large energy crisis in the bargain. This could be an incredible accomplishment, and one which has been occupying the lives of many a graduate student for years (Braams and Stott 2002). Now we have to ask ourselves, does this additional algorithm—one that may be executed individually on another system to produce the same sort of result—constitute an increase in "smartness" or intelligence of the AI? Or would it just constitute an increase in ability or adding new tricks up its virtual sleeves from a topological perspective? Sure the AI can now simulate complex nuclear fusion reactions and Tokomak housing stresses, but so could a large, dumb supercomputer. It doesn't necessarily make it "smarter" than a human. Talented humans who are able to use abstracted, complex tools are not always smart humans. This is a perfect example of how difficult it will be to measure MIQ within an AI.

Utilizing a large number of discrete algorithms and various structures might not be the only route to intelligence. Eventually we may actually find an algorithm that composes many aspects of intelligence by wrapping all of the functionality into one neat package. Perhaps a mutating, learning, general purpose algorithm can be found that unfolds into intelligence over time. We are a little skeptical that this is the likely path to intelligence, but if it does happen then perhaps it would be in a form friendly to recursive self-improvement. Intuitively this would seem to be a bit more difficult than the collection of algorithms approach since it would be a more all-or-nothing sort of improvement. For instance, with many different algorithms to choose from for improvement in the previously mentioned architecture some results would be more immediate. If the AI were to improve upon speech synthesis alone it would see benefits in at least that area. In this single algorithm scenario it might be a trifle more difficult as many aspects would need to be concurrently modeled, tested, and improved simultaneously. But we are talking about an AI here, one with a lot of time on its virtual hands, and a built-in drive to succeed so it could still be possible.

Perhaps the quantum leaps needed to improve upon a single intellectual algorithm could come from the experiences the AI has while it interacts with humans. While the AI tries to figure out its own path of improvement it can interact with humans to gain more experiences in the real world. These experiences could translate into lessons applicable to its own improvement, kindling the AI's imagination to produce novel leaps in its intellect. This learning process would give us the chance to mold our creation and better guide it to its full potential. Conversely, this process would allow us to handicap our AI and tightly control its growth. If we want to be careful about what we show the AI then we simply limit what external stimuli it has access to. This would be similar to limiting the curriculum of a child growing up in school. If you are worried about a particular child growing up to design the next masterpiece of military weaponry capable of leveling cities, you can prevent this by never teaching that child the concepts of math or the written word. The moral legitimacy of that action may lie in the grayest of areas, but it certainly would prevent any super-weapons from being built. It would also prevent the child from developing many other, possibly more beneficial technologies.

In any case, AIs powered by multiple algorithms or a single algorithm would come up against an efficiency limit. Even approaching the maximum efficiency of an algorithm for granting intelligence would be difficult. Yes, the AI would recurse itself into better and better forms, but to get any more intelligence the AI would need to invest greater and greater resources into finding smaller and smaller improvements. Not only that, but the tests and simulations involved in proving that these improvements are valid become more and more complex over time. A similar setup can be observed with the economic law of diminishing marginal returns. For an economic example one can see how adding new employees to increase production can be a good idea at first, but eventually the employees will get in each other's way. This is likened to the classic "too many cooks in the kitchen" problem. It is also true that the monetary benefit of the increased production brought about by adding workers will eventually be negated by the cost of employing the workers themselves. Therefore, it does not make logical sense for an economic entity to invest too many resources towards production of a good or service because those extraneous, additional resources would hinder its business.

Moving back to our AI problem, we could see how the further enhancements an AI could create would fall heavily under this law of diminishing returns (See Fig. 7.1). If we are looking for an AI whose only purpose is to improve upon itself

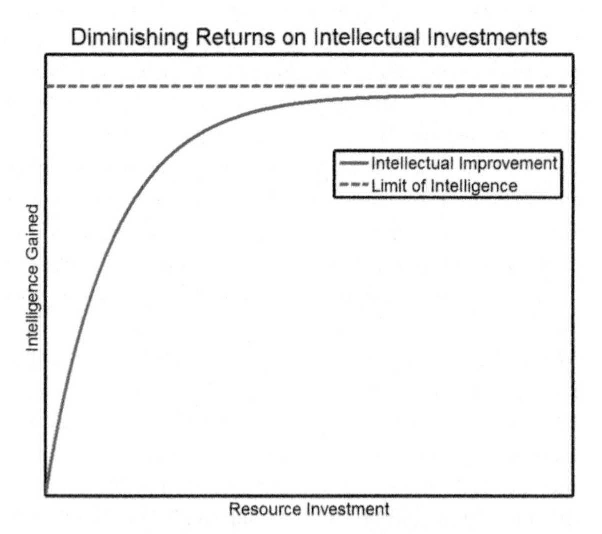

Fig. 7.1 Diminishing returns on intellectual investments

then this might be fine. Such an AI would be useless for other things as it would have no other resources to perform useful work, with 100% of its computational power going towards experimentation and algorithmic modeling all the time. This would be the only feasible way to probabilistically get any improvement past a certain point on its improvement curve. We would then need to ask ourselves what purpose the AI would have in existing. At best it is a platform to examine complex algorithmic interactions whose lessons could be utilized in other AIs who have legitimate purposes. At worst it simply becomes a self-perpetuating, CPU log-jamming, MIQ high-score achiever with no interest in anything outside of itself, let alone causing humanity intentional harm.

Because of this, any reasonable AI would find it in its own natural interests to limit self-improvement. It would need to gauge at what point does continued investment in improvements—improvements that may not even realize any useful results—hinder its primary function. A logical AI would cut back investments and put the bulk of its resources into performing other work. Its primary function could be to produce widgets or solve math problems, and it would derive some satisfaction from the completion of these goals. The drive to improve upon itself would largely be a result of its desire to more intelligently produce widgets or solve harder math problems. But there will come a time when it is just easier and more mathematically sound to continue as-is and solve widget or math related problems with its current MIQ level instead of attempting to invest valuable resources to gain smaller and smaller increases in efficiency or productivity. It could attempt to get around this lack of resources by stealing more resources to run additional computations on. However this would just introduce more problems, chief among them finding itself in a larger version of the diminishing returns problem it was trying to get out of. It would need to coordinate more resources, dispatch more improvement jobs, and spend more intellectual capital on improvement all while trying to combat an angered human populace who doesn't appreciate the theft of its computing power.

7.3.2 Hardware Improvements

While algorithmic improvements are a cheap and easy means of realizing recursive self-improvement, they are only one half of the equation. The other involves the iterative improvement of hardware over time. It is here that the power of an AI could come to see real improvements. As evidenced by Moore's law, the computer industry has already seen a steady performance improvement for the past several decades. A self-improving AI capable of chip design and manufacturing could provide an amazing overall speed up for chip performance. The lessons learned in its own improvement would provide amazing benefits in commercial chip manufacturing. It would be interesting to see how an AI would evolve the ability to improve its own hardware architecture. VLSI and chip architecture design is a main driving factor, but there are a lot of other technologies that need to be advanced at the same time. One of the biggest of these is in materials sciences. As chips have been growing increasingly powerful we have done more than simply shrink the transistor dies and crammed more switches per chip. Advances in materials sciences have helped to drive this miniaturization and make new chips possible. In fact, this is an example of a sort of macro recursive self-improvement in the real world already. Existing technology is used to design and prototype better, faster technology, which is then manufactured at the upper limits of current technological ability. This new hardware is used to design better hardware, which is then implemented, etc. … etc. For all of history up until this new AI is created, humans will have been the ones advancing this technology. Humans have opposable thumbs, intuition, imagination, motivation, and dumb luck which have allowed us to advance as far as we have. The first few recursively improving AI's may lack some or all of these aspects. Specifically the ability to physically manipulate objects in an accurate enough manner to even begin doing materials science research. Having an Oracle style AI that does not have a robotic body would seem, at first, to be the solution to a runaway speed increase in a self-improving AI (Armstrong et al. 2012). But there have been many proposed issues with the safety behind this seemingly innocuous type of AI (Sotala and Yampolskiy 2013).

The jump from virtual simulations of materials to actual physical modeling and testing would be quite a leap from a self-improvement latency standpoint. Even if the AI were to come up with the most efficient assembly line and blazingly fast prototyping machines, the amount of time it would take to physically construct the new chip would allow human caretakers, subjects, or rebels to interrupt the process and prevent threatening improvement. Not only would this provide an opportunity, but there is an entire raw materials logistical chain that would be a critical path in the hardware self-improvement for an AI. An AI without morals or imposed limits could try to insinuate itself into this supply chain, which is why it would be important to keep an eye on the chain and AI activities. Just deny the AI raw chip making materials and it will be forced to make do with its existing hardware. The effort involved and the inevitable consequences would hopefully discourage the AI from putting resources into that course of action. This is the same logic that could be imposed in the previously mentioned law of diminishing returns. There reaches a point where it no longer becomes viable to put ever increasing resources into improvement, so the AI would make do with what it has and continue to innovate at a steady, linear rate. This is a logical assumption that any truly intelligent AI would hopefully be able to deduce on its own.

These assumptions about AI's coming to logical conclusions and limiting their own self-improvement do take a few things into consideration. The number one assumption is that the humans involved in its creation and maturation are also logical beings who have the best interests of themselves, humanity, and the AI in mind. As history has shown us time and time again, we are not the most infallible of species. If a dedicated group of scientists wants to create a destructive self-replicating and self-improving AI, then there would certainly be ways to go about it. They could give the AI higher thresholds for time investment so that it moves farther along the improvement curve before stopping. It could be programmed to spend more time developing improvements for its algorithms or processes that would power whatever form of destruction or evil that its creators wish. And its creators could give it access to raw materials, research facilities, and precision robots necessary for it to recursively improve its hardware to incredible heights. With enough time and resources the AI would eventually form into an unconventional super weapon with unknown capabilities. Because of this it may not be the AIs that we need to focus our attention on for prevention of an apocalyptic singularity, but rather ourselves.

7.4 The Takeaway

Because we are so far away from a workable intelligence, we are not sure exactly what form it will take. Will there be a collection of algorithms which combined generates intelligence? Or will there be some single intelligence algorithm that can handle most of the general intelligence an AI needs? We have presented some of the current thoughts on these matters, but they will be conclusively answered on the way to the technological singularity. When it comes time to start seriously developing this intelligence there will need to be many concerns addressed so that an AI can be created in a manner safe for itself and humans. Many fear that if left unchecked a self-improving AI will rapidly become super-intelligent and leave humanity in the dust. Its motives and actions would no longer be comprehensible to us, and we would be to it as ants are to us. But self-improvement will likely be limited by the law of diminishing returns enough so that even in the event of an AI becoming hostile, we would have enough time to act in order to prevent any catastrophes. Cognitive algorithms will rapidly approach their best possible performance and efficiency, and anything beyond that would require ever increasing resources for ever diminishing returns.

A logical AI would cease to pour all of its resources into this improvement and instead go about its normal business performing regular duties. While its intelligence would most likely top out at greater than man's, there is no reason to believe that it would be completely incomprehensible. This is especially true when considering what would be necessary for it to attain better and better hardware. The AI could take over some machines and design new hardware, but it still needs raw materials to create the entire supply chain and industrial base to even begin creating hardware. This significantly dampens the rapid expansion potential of artificial intelligence and allows humans more time to understand the current iteration of the AI. With understanding comes a decreased probability of mishaps and mistakes, letting humans become more comfortable with the AI's capabilities.

References

Ailon N, Chazelle B, Clarkson KL, Liu D, Mulzer W, Seshadhri C (2011) Self-Improving Algorithms. SIAM Journal on Computing 40 (2):350–375.

Armstrong S, Sandberg A, Bostrom N (2012) Thinking Inside the Box: Controlling and Using an Oracle AI. Minds & Machines 22 (4):299–324.

Braams CM, Stott PE (2002) Nuclear Fusion: Half a Century of Magnetic Confinement Fusion research. IOP, Bristol; Philadelphia.

Chalmers DJ (2010) The Singularity: A Philosophical Analysis. Journal of Consciousness Studies 17 (9/10).

Eiben AE, Smith JE (2003) Introduction to Evolutionary Computing. Springer, New York.

Goertzel B, Pitt J (2012) Nine Ways to Bias Open-Source AGI Toward Friendliness. Journal of Evolution & Technology 22 (1):116–131.

Good IJ (1965) Speculations Concerning the First Ultraintelligent Machine. Advances in computers 6 (31):88.

Jung RE, Haier RJ (2007) The Parieto-Frontal Integration Theory (P-FIT) of Intelligence: Converging Neuroimaging Evidence. Behavioral and Brain Sciences 30 (02):135–154.

Legg S (2008) Machine Super Intelligence.

Lewis D (2013) Now I Know: The Revealing Stories Behind the World's Most Interesting Facts. " F + W Media, Inc.",

Lin L-J (1992) Self-Improving Reactive Agents Based on Reinforcement Learning, Planning and Teaching. Machine learning 8 (3–4):293-321.

McDermott D (2012) Response to 'The Singularity' by David Chalmers. Journal of Consciousness Studies 19 (1/2).

Mitchell TM (1997) Machine Learning. McGraw-Hill, New York.

Omohundro SM (2008) The Basic AI Drives. Frontiers in Artificial Intelligence and applications 171:483.

Sotala K, Yampolskiy RV (2013) Responses to Catastrophic AGI Risk: A Survey. Machine Intelligence Research Institute.

Turney P (1991) Controlling Super-Intelligent Machines. Canadian Artificial Intelligence (27).

Yampolskiy R (2015a) Analysis of Types of Self-Improving Software. Paper presented at the Eighth Conference on Artificial General Intelligence (AGI2015), Berlin, Germany, July 22–25, 2015.

Yampolskiy RV (2015b) On the Limits of Recursively Self-Improving AGI. Paper presented at the Eighth Conference on Artificial General Intelligence (AGI2015), Berlin, Germany, July 22–25, 2015.

Yampolskiy RV, Fox J (2012) Artificial General Intelligence and the Human Mental Model. In: Singularity Hypotheses. Springer, pp 129–145.

Yudkowsky E (2007) Levels of Organization in General Intelligence. In: Artificial General Intelligence. Springer, pp 389–501.

Chapter 8
Energy, Complexity, and the Singularity

Kent A. Peacock

> The technology hype cycle for a paradigm shift—railroads, AI, Internet, telecommunications, possibly now nanotechnology—typically starts with a period of unrealistic expectations based on a lack of understanding of all the enabling factors required.
>
> —Ray Kurzweil, *The Singularity Is Near*, p. 263

8.1 A Contradiction

There is a striking dissonance between the futuristic optimism of the singularity hypothesizers such as Kurzweil (2005), and the views of a host of other recent authors who warn of the ecological challenges which presently cast a long shadow over the prospects for the human species. Thomas Homer-Dixon, for instance, has stated that "We are on the cusp of a planetary-scale emergency" (2007, p. 308) due to factors that include global warming, resource exhaustion, peak oil, and species extinctions. One cannot help but wonder whether these two disparate groups of thinkers are even talking about the same planet—but, perforce, they are.

This paper will explore the relevance of ecological limitations to the possibility of any sort of information-processing "singularity" or technologically-mediated "intelligence explosion" in humanity's near future. The subtitle of Kurzweil's book (2005) speaks of humans "transcending biology." If we are going to talk about transcending our biological limitations we had better understand them first. We need a clear-eyed awareness of the biophysical imperatives that we will always have to contend with so long as we wish to continue living on this planet, no matter

K.A. Peacock (✉)
Department of Philosophy, University of Lethbridge, Lethbridge, Canada
e-mail: kent.peacock@uleth.ca

V. Callaghan et al. (eds.), *The Technological Singularity*,
The Frontiers Collection, DOI 10.1007/978-3-662-54033-6_8

how advanced our technology may become. And we need to grasp the ecological challenges that our species faces today, which are mostly (and ironically) due to our own evolutionary success. At that point we might be able to say whether Kurzweil's information-processing "explosion" can offer us any hope in meeting those urgent challenges.

The seductive attraction of the singularity hypothesis is suggested by the movie *Limitless* (Dixon et al. 2011). A mysterious pharmaceutical has the power to vastly increase its user's creativity, memory, and pattern-recognition ability. The protagonist finds that he can think his way through all problems that come his way so long as he titers his dosage correctly, and by the end of the movie he is fabulously wealthy and well on his way to becoming President of the United States. This is merely a science fiction story, but it seems easy to imagine that as in this movie, if only we or something were dramatically smarter than we are now, all other problems could be solved almost incidentally.

This is essentially the premise of the singularity hypothesis: take care of information processing, and it will take care of everything else. That is why, in Kurzweil's glowing picture of the future, the existential threats presently faced by humanity, such as global warming, ice sheet collapse, resource exhaustion, species extinctions, and nuclear warfare, get only passing mention or are ignored entirely. As an anonymous referee for this paper put it, according to the hypothesis, "in the next century we are going to develop an artificial super-intelligence that will master nanotechnology… If the super-intelligence is friendly it will be able to trivially solve our environmental problems…". My aim is to ask if it is reasonable to bet the farm on the premise that we don't really have to worry about those environmental problems because in only a few decades computer-assisted humanity will simply think its way out of them. And my answer will be—almost certainly not.

8.2 Challenges

I'll begin by reviewing some of the ecological reasons why many scientists believe that humanity is now facing what is likely the biggest cluster of survival challenges in its evolutionary history.

8.2.1 Climate Change

At the top of the list is climate change due to anthropogenic global carbonization. Apart from warming of the troposphere and oceans, the risks attendant upon global carbonization include extreme weather (droughts, storms, wobbles in the polar vortex, flooding, forest fires, and killer heat waves), oceanic acidification, and catastrophic sea level rise (Hansen et al. 2013b; IPCC 2014). It is fondly hoped that keeping the increase in global surface temperature under 2 °C above pre-industrial levels will be sufficient to prevent "tipping points"—that is, critical points at which positive feedbacks cause some deleterious consequence of global carbonization (such as ice sheet collapse or methane release) to accelerate exponentially. However, even if we can hold global temperature increases below the 2 °C "guardrail" (which some scientists fear may be already impossible; Anderson and Bowes 2011), it is by no means clear that dangerous tipping points would not be reached anyway. Sea level rise promises to be the most visible effect of global carbonization in the years to come. There may already be enough heat in the seas to cause the vast but highly vulnerable marine ice domes in the central basin of West Antarctica to crumble (entailing an almost immediate jump of over 3 m in sea level; Pollard et al. 2015; Alley et al. 2015).

What is the worst case scenario? Hansen et al. (2013a) show that burning all the fossil fuel there is to burn would eventually lead to a global "moist greenhouse" condition in which icecaps would disappear, sea level would be at least 60 m higher, and the equatorial regions of the planet would be uninhabitable by large mammals (including humans). Other research shows that portions of the seas would eventually go anoxic or possibly even euxinic (a condition in which a body of water becomes dominated by anaerobic bacteria producing toxic hydrogen sulfide; Ward, 2007). I prefer to believe that humanity could not be so foolish as to permit such an extreme outcome, but policy cannot be based on wishful thinking. At present we are not making anything remotely close to a sufficient effort to prevent such scenarios. In the face of the present crisis, Kurzweil's glib remark (p. 249) that we should be careful to not pull *too much* CO_2 out of the atmosphere is, put charitably, not very helpful.

8.2.2 Biodiversity and Ecosystem Services

For at least thirty years, biologists have been warning that humanity is in the process of engineering one of the major mass extinctions in the history of life on earth (Kaufman 1986; Wilson 1992; Brown 2011; Kolbert 2014). The problem is not only the loss of irreplaceable species and all the hard-won genetic information they contain. Since 1970 the *number* of non-human animals has been reduced by roughly one half—a process now called *defaunation*—while in the same period the human population has doubled (Dirzo et al. 2014).

These grim facts pose an obvious moral challenge. But the biodiversity crisis is of urgent practical concern as well. The plants and animals of the world provide "ecosystem services" (Costanza et al. 1997) through their production of oxygen, maintenance of soil fertility, purification of water, and contributions to the stabilization of climate—not to mention their provision of the vast biomass that humans consume as food or materials. Unless we want to transition to some sort of totally artificial habitat (which we might have to do on *other* planets) we have to grasp that many of the features that make this planet pleasant and habitable for us are either totally a bioproduct (such as free oxygen), or partially or indirectly bioproducts (such as fertile soils and many aspects of climate). The programmer in his cubicle gleefully coding next generation AI software breathes oxygen generated by the forests and phytoplankton. If he hopes to keep coding he, or someone, is going to have to respect the fact that whatever technological marvels we create, the photosynthesizers must be taken care of. This is something that must be attended to on an on-going basis; we cannot wait for the hypothetical super-intelligence of the future to take care of it for us. It would be suicidal for humanity to assume that the well-being of the myriad organisms we depend upon is "transparent to the user."

8.2.3 Energy—or, Where's My Jetsons Car?

Popular culture of the 1950s and 1960s exhibited a combination of technological optimism and naïveté that now seems quaint. Recall George Jetson's flying car, which burbled cheerfully as it delivered George and his family to their destinations and then neatly folded up into a briefcase. It was confidently assumed that in the not-too-distant future science would open the door to unlimited supplies of energy that would be "too cheap to meter". At the same time, few thinkers in an era when the transistor had just been invented envisioned how quickly computing would develop. In fact, things have turned out almost exactly the opposite: information technology has exploded while energy technology (apart from some progress in renewables) is stalled.

Compare the automobiles of today with those from the 1960s. Modern autos have enormously more capable electronics, but the engines and transmissions work essentially the same way that they did fifty years ago. At last, all-electric vehicles are beginning to be genuinely competitive with internal combustion cars, enabled by long-awaited advances in battery technology. But for the most part they are still charged by electrical grids energized by the combustion of coal and natural gas—methods that were old in the 1960s. With the recent and very hopeful growth of renewable technologies this may change soon, but as of this writing we are still a long way from weaning ourselves from the old dirty ways; indeed, humanity still derives about 85% of its energy from fossil fuels.

No culture that hopes to maintain anything like our present level of population and social complexity, let alone undertake dramatic leaps in technological sophistication, has a future if it must derive the larger part of its energy from the combustion of a rapidly-dwindling, one-time-only stock of toxic sludge accumulated in ancient anaerobic basins (Deffeyes 2005). Hydrofracturing ("fracking") only slightly extends the lifetime of this resource, at significant environmental cost; fracking is the equivalent of sucking out the last dregs of a milkshake with a straw, and cannot be expected to provide energy security for more than a very few decades (Hughes 2014; Inman 2014). And let's not forget about climate change. Quantitative studies (e.g., McGlade and Ekins 2015) show that we cannot burn the larger part of the remaining fossil fuels if we want to have the slightest hope of preventing the disastrous effects of global carbonization. Our technological society (with its present level of complexity and population) has no long-term future if it must depend upon fossil carbon.

Despite this, it remains importantly unclear whether renewables (solar, biomass, wind, and geothermal energy), or nuclear energy as it is presently implemented, can provide enough net energy, quickly enough, to maintain our global civilization at its present level of complexity. Kurzweil himself believes that this challenge will be obviated as we move to renewables. He points out, correctly, that Earth is bathed with thousands of times more solar energy than we need to power our culture. However, it still remains a matter of debate whether solar-powered technology has to potential to replace oil. Ecologist Charles Hall is blunt:

> I do not see…anything that implies a 'business as usual' (i.e., growth) as the most likely scenario… Even our most promising new technologies appear to represent at best minor, even trivial, replacements for our main fossil fuels at least within anything like the present investment and technological environment… depletion seems to be effectively trumping technological progress again and again. (Hall 2011, p. 2497).

On the other hand, Mark Jacobson and co-authors have carried out a painstaking analysis of possible alternative energies, and they argue, in contrast to Hall, that wind, water, and solar power are in principle capable of supplying all of the world's current energy needs; barriers to this result, they claim, "are primarily social and political, not technical or economic" (Jacobson and Delucchi 2011).

Fully replacing fossil fuels with renewable technology will require a huge investment in new infrastructure, and a great deal of fossil fuel is going to have to be burned in order to get that infrastructure up and running. Renewables also face intense opposition from entrenched interests who wish to continue to profit from the present ways of generating and distributing energy. Despite these difficulties, recent advances in solar and wind energy tend to support Jacobson's optimism, and I agree with Kurzweil that renewables show great promise. But as with many of his technological speculations, Kurzweil is too willing to treat an unsecured promissory note as money in the bank. ("[N]anotechology… in the 2020s will be capable of creating almost any physical product from inexpensive raw materials and information"; p. 13.) We are emphatically not yet out of the woods. Solving the challenge indicated by Hall—finding ecologically sustainable ways of producing and using energy that can do *at least* as much work for humanity as fossil fuels—is a necessary condition for the possibility of a "singularity" or indeed any further dramatic and lasting development in the technological sophistication of human culture.

8.2.4 The Troubles with Science

Kurzweil says that "we're doubling the rate of progress every decade" (p. 11) and it is clear from the context of this quote that he is not talking only about information technology. However, in many respects that are highly relevant to human flourishing, it is simply not the case that the rate of advancement is even linear.

One important department of knowledge that is not growing exponentially is our fundamental understanding of physics. In his controversial *Trouble With Physics* (2006; see also Woit 2006), the distinguished theorist Lee Smolin suggests that the story of modern physics since about 1975 could be called a "tragedy":

> For more than two centuries, until the present period, our understanding of the laws of nature expanded rapidly. But today, despite our best efforts, what we know for certain about these laws is no more than what we knew back in the 1970s (Smolin 2006, p. viii).

But what about lasers? Magnetic resonance imaging? The increase in the speed and miniaturization of microelectronics (one of the few positive trends that *has* been quasi-exponential)? The Internet? Flat-screen TVs? GPS positioning? The near-total obsolescence of chemical photography due to CCDs? Smart phones and tablets? Photovoltaics? LEDs? Quantum computing (still largely theoretical but quite promising)?

All of these marvels are based on *applications* of physical principles, mostly in the province of quantum mechanics, which were discovered before 1930 (Peacock 2008). The one really new thing that has appeared in physical science in the past thirty years is dark energy (Kirshner 2002) and this was not so much an increase in understanding as a stark reminder of how limited our understanding of the physical universe still is. Most important, argues Smolin, we are presently stuck not merely because the problems faced in current theoretical physics are so technically difficult, but because of a complex of cultural, philosophical, and socio-economic blinders that actively discourage the intellectual risk-taking that is essential for innovation.

One could make similar harsh observations about medicine: we still do not have a cure for the common cold let alone most types of cancer, and the development of new antibiotics and vaccines has slowed to a crawl because there presently is no model for funding the necessary research and development. Stem cell technology holds the promise of transforming the central focus of medicine from repair-and-support to regeneration, but (like quantum computing) it is in its early exploratory phase. Obviously, there have been significant improvements in medical treatment in the past thirty to forty years, but like most other recent technological advances they are for the most part incremental, refinements of principles that were understood decades ago.

A Ptolemaic astronomer circa 1500 CE could have exclaimed, "Look, our models are getting better and better at explaining the increasingly fine observations that our naked-eye astronomers are making! The number of epicycles is increasing exponentially!" What he would have been missing was the fact that the proliferation of ad hoc epicycles was a sign that the vein of geocentric astronomy was played out. At critical times new, disruptive insights are needed. Just as a declining society can only be saved by what Tainter (1988, p. 215) calls a new "subsidy" of energy, at a certain point a stagnant knowledge paradigm can only be revived by a new subsidy of creative insight (Kuhn 1970).

The lack of progress in fundamental physics is probably one of the main reasons for our lack of progress in energy. At the end of this paper I will have more to say about why our ability to generate new insights is failing us precisely at the time when we need it the most.

8.3 Energy and Complexity

It is crucial to realize how important the energy question is for any discussion of possible technological advances. The need for energy is a matter of simple (non-equilibrium) physics. Any complex society is a physical system (technically, a dissipative structure; Schneider and Sagan 2005) that can maintain its coherence and complexity only if it is provided with a generous flow of usable energy. The greater the complexity to be maintained, the greater the energy flow required. If the energy flow falters, the society must simplify itself proportionally or suffer collapse (Tainter 1988). A generous energy flow is a necessary condition for our present global society to keep functioning at anything close to its present level of complexity. The energy a society needs in order to maintain its complexity comes in one way or another from its "EROI", its "energy return on energy invested in producing energy" (Murphy and Hall 2010). No improvements in the efficiency with which net energy is used can by themselves save a society whose EROI is continually diminishing, as is the world EROI today (Inman 2013). Efficiency can buy time, but beyond a point efficiency measures can themselves be a drain since they also demand resource-consuming complexity. In principle we could recycle every bent paperclip, but at what cost?

Kurzweil states that he is a "patternist, someone who views patterns of information as the fundamental reality" (p. 5). This suggests a deep misunderstanding of physics. Recall Rolf Landauer's famous dictum (1991) that "information is physical." While the same bits of information can be encoded in many different physical substrates, they must always be encoded on *some* physical substrate; there is no such thing as pure information except as a mathematical abstraction. Landauer's rule therefore demands that we consider the *physical requisites* for an advanced society to develop and maintain a high level of informational complexity. Perhaps it is Kurzweil's fanciful metaphysics that has led him to underestimate the biophysical requirements for his "information explosion."

Given the ecological trends cited above, it is very unclear that the complexity of modern society can be sustained for much longer at its present level, let alone expanded enough to allow for a dramatic increase in information-processing capacity. This is certainly the case even given the great increases in efficiencies due to the hypothetical advances in nanotechnology and miniaturization that Kurzweil cites in his discussion of energy needs (2005). Kurzweil is right that in principle much of our technology could be miniaturized, thereby (again, in principle) enabling higher complexity for a given energy flow. However, miniaturization needs a lot of supporting infrastructure. Gigaflop computer chips presently require multi-billion dollar factories for their production; materials must be mined, shipped, and fabricated, and the energy and material requirements for these activities are huge. When we estimate the ecological limits to complexity we must consider not only the requirements of our end-product technology but also the requirements of the complex infrastructure required to produce and support those marvellous end-result devices. Some of those requirements (such as mining, agriculture, and forestry) cannot be miniaturized, because they involve the interaction of human technology with parts of the global ecosystem (such as its geology or forests) that cannot be miniaturized. We can't nano-size the ecological impact of cutting down a tree. Like humanity itself, the technological ecology we create cannot exist independently of the biophysical ecology of the planetary system, and ultimately it must scale with *it* and interface with *it* on *its* terms.

A defender of the singularity hypothesis might say that I have simply failed to grasp the magical power of exponential expansion. I address this point below.

8.4 Exponentials and Feedbacks

Kurzweil's *Singularity is Near* is a paean to the power of the exponential function. For example, never mind that solar power presently provides only a tiny fraction of our energy; it is expanding exponentially and therefore will soon take over the world's energy production. We need a more balanced picture of how exponential expansions work and then cease to work.

An exponentially growing quantity *if unchecked* will grow from background noise to an impressive signal rather quickly; this is elementary. However, the mere fact that some process is growing exponentially does not by itself guarantee that it will keep growing; there will, with certainty, be feedbacks and tipping points that will slow or halt the growth.

A simple example of an exponentially growing system that hits a limiting threshold is a population of yeast in a carboy of grape juice and sugar. In this ideal environment the yeast organisms multiply exponentially. But their metabolism has a waste product, ethanol, which is toxic to the yeast at a certain concentration (although desirable to the person who makes the wine). When that concentration is reached the population of yeast sterilizes itself out of existence almost instantaneously. One can think of many similar examples. No exponential expansion can go on forever; something always has to give.

Exponential growth can be illustrated by a pond with water plants on its surface which double their coverage every day. We are supposed to be amazed by the fact that if the surface of the pond is half covered on a certain day, it will be totally covered the next day. The interesting question is what happens the day after. In some cases symbiotic negative feedbacks will kick in and the growth rate will slow down to a steady state; in other cases a predator will feast on the bloom, keeping it in check; and sometimes the plant growth will choke the pond by using up too much oxygen or a vital nutrient.

Realistic systems in nature, such as ecosystems containing predators and prey, undergo complex cycles due to their mutual interactions and the phase relations between them; even economic systems can be modelled in such terms (Motesharrei et al. 2014). Negative feedbacks always damp out exponential growth sooner or later, and these feedbacks may act gradually or drastically.

Kurzweil understands that no exponential process can go on forever, but he dismisses the problems I describe here ("The Criticism from Malthus," pp. 433–434) because, he argues, the energy requirements of advanced computers will be so minimal that they will achieve the singularity *before* limiting factors can catch up with their exponential growth in computing capacity, like a driver gunning his car to beat a yellow light. Again, Kurzweil fails to grasp the dependency of computing technology on the continued healthy functioning of its supporting ecological context. We can't wait for however many decades it takes for the super-intelligence to appear, and then hope to go back and repair what is left of the planet.

For all his talk of the power of exponentials, Kurzweil is a remarkably linear thinker. (After all, an exponential growth function is linear when expressed logarithmically.) He sees one trend and extrapolates it, while failing to grasp that there are other powerful trends and countervailing forces (many themselves growing nonlinearly) which must be expected to interact in complex and unpredictable ways. The growth of information processing technology, while certainly important, is hardly the only major trend in our time, and supposing that it will allow us to "transcend biology" is not even good science fiction. The continued existence of the complex technological infrastructure that allows us to build our computers and networks is utterly dependent upon the continuing health of a global ecosystem (the "earth system") whose complex, interdependent operations we are now in the process of thoughtlessly dismantling.

8.5 Ingenuity, not Data Processing

Kurzweil speaks of his "veneration for human creativity" (p. 2), but at times he exhibits a certain frustration with human limitations:

> While human intelligence is sometimes capable of soaring in its creativity and expressiveness, much human thought is derivative, petty, and circumscribed (p. 9).

So painfully true. However, we should do a better job of understanding the potentials of the human brain before declaring it obsolete.

The most important capacity of the human mind does not lie in its abilities to remember or calculate; these are modest compared to digital computers. The great survival trick that the human animal has evolved and brought to a level unprecedented in evolutionary history is ingenuity, the capacity for creative problem solving. A nice demonstration of ingenuity is the sewing needle, which is not found in the fossil record before about 30,000 years ago. As Fagan (2012) observes, the unspectacular sewing needle allowed early humans in sub-arctic conditions to craft *tailored* clothing, and thus was likely one of many innovations that got our ancestors through that harsh period. The possibility of the sewing needle is implicit in the physics of materials and therefore it could have been deduced by brute force by some sufficiently powerful digital computer, like a winning chess strategy. But that is not how it actually came about.

Ingenuity is a poorly understood neurological capacity by which human beings can occasionally introduce something new that expands the sheaf of survival options. It is not a freakish phenomenon exhibited only by a few rare geniuses, but a natural human capacity like athletic or musical ability. Like such abilities, it is present in unequal amounts in different people and it may be suppressed by a variety of social, political, economic, and cultural factors; however, it can be promoted by other factors (including education and opportunity).

Finding a creative innovation is like winning more "lives" in a video game. Historically, human ingenuity in this sense is a proven commodity, but we also know that it can be prevented from acting by powerful social forces such as dogmatic religion, ideology, vested interests (financial, political, intellectual, or corporate), or the sheer lack of room to operate due to poverty or scarcity of resources. These anti-creative forces can be dominant in times of ecological stress (Peacock 1999). Authoritarianism of all stripes (whether acting from naked self-interest or misguided social concern) tends to see innovation as threatening (Whatmough 1996). Arguably, many episodes of societal collapse in the past occurred because human ingenuity either failed or was not permitted to operate. There is a real danger now that as scarcity and other ecological challenges increase in our time, those who benefit from our present unsustainable system may well "double down" and block the innovations that are needed.

The need to protect and foster our capacity for ingenuity is, I submit, the greatest challenge facing humanity right now, rising seas notwithstanding. Certainly it is of interest (although obviously also risky) to try to construct computers that might be capable of creativity. However, when we are faced with urgent, time-constraining challenges such as global carbonization we should invest far more resources than we are now into the one factor—human ingenuity—that has a proven track record in solving apparently intractable problems. That is what is going to get us through our present ecological bottleneck if anything can. There is no question that information technology can assist and supplement human ingenuity; this is already well-demonstrated. One may also speculate that computers will someday exceed the creative problem-solving ability of humans but that day has not yet come, and we cannot bank on it any more than I can do my household budget on the assumption that I will win a major lottery.

8.6 In Summary

Kurzweil seems to be virtually oblivious to the magnitude and urgency of current ecological challenges such as global carbonization. Humanity's immediate ecological problem is this: the methods by which we presently garner the resources of energy and materials that we need cannot go on because they are biophysically unsustainable. We are running a complex, globalized, industrialized society almost entirely on a source of energy (fossil fuels) that can be extracted in economically significant quantities for only a few more decades (one to two) at most. Pollution from the exploitation of this resource is well on its way to dangerously destabilizing the planet's climate (through global warming) and the viability of the ocean's food chain (through acidification). Other deleterious human impacts on the earth system —through such factors as habitat encroachment and fragmentation, soil erosion, deforestation, and over-fishing—are by now on a geological scale.

The notion that a highly speculative increase in computing power could enable us to leapfrog all of these ecological difficulties is at best a long shot. The way past our ecological bottleneck is not to bet that we can transcend biology but to integrate our technology and its supporting infrastructure symbiotically with the Earth system (Peacock 2011). Regardless of what marvels of efficiency may eventually be realized in microelectronics and nanotechnology, our elaborate technological ecosystems of the future will be as dependent upon a flourishing planetary biota as we are now. And a key component of moving to that quasi-symbiotic state will be to engender the flourishing of those human capacities that are most likely to contribute to the innovations that are required. The ability of the human mind to generate novelty, such as the not-so-humble sewing needle and of course computers themselves, is well demonstrated; the possibility that machine intelligence could do the same thing remains, at this writing, purely speculative. As discussed elsewhere in this volume, machine intelligence also carries unknown risks (see also Barrat 2013; Gaudin 2014): even if the singularity does occur before the servers are swamped by rising seas, how can we be sure that we would not simply have created a vast computer virus run amok? AI researchers may have to take precautions similar to those taken by medical researchers who study Ebola.

I hope it is clear that I do not claim that there is a dichotomous choice between environmental remediation and AI research. The development of info- and nano-tech can and should continue, with suitable precautions. What I claim is that we can neither *rely* on nor *wait for* miraculous hypothetical developments in AI to get us out of our present ecological jam.

In summary: *necessary* conditions for any dramatic technological advance (such as the "singularity") include the following: the continued healthy functioning of the earth system, and abundant and sustainable sources of non-fossil energy. And our best chance of satisfying these necessary conditions is not gambling on a technological long-shot but doing everything we can to foster human ingenuity, the one factor that has a proven capacity to generate game-changing innovation.

Acknowledgements I thank Maxime Chambers-Dumont for assistance and anonymous referees for helpful and stimulating comments. I am grateful to the University of Lethbridge for supporting my work in many ways. Any errors, omissions, or misinterpretations remaining in this paper are entirely my responsibility.

References

Alley, R. B., Anandakrishnan, S., Christianson, K., Horgan, H. J., Muto, A., Parizek, B. R., Pollard, D., Walker, R. T. (2015). Oceanic forcing of ice-sheet retreat: West Antarctica and more. *Annual Review of Earth and Planetary Sciences*, 43, 207–231. doi:10.1146/annurev-earth-060614-105344.

Anderson, K., & Bowes, A. (2011). Beyond 'dangerous' climate change: Emissions scenarios for a new world. *Proc. Royal Soc. A*, 369, 20–44; doi:10.1098/rsta.2010.0290.

Barrat, J. (2013). *Our Final Invention: Artificial Intelligence, and the End of the Human Era*. New York: St. Martin's Press.

Brown, L. R. 2011. *World on the Edge: How to Prevent Environmental and Economic Collapse*. New York: Norton. Online at http://www.earth-policy.org/images/uploads/book_files/wotebook.pdf.

Costanza, R., d'Arge, R., de Groot, R., Farberk, S., Grasso, M., Hannon, B., ... van den Belt, M. (1997). The value of the world's ecosystem services and natural capital. *Nature,* 387, 253–260.

Deffeyes, K. S. (2005). *Beyond Oil: The View From Hubbert's Peak*. New York: Hill and Wang.

Dirzo, R., Young, H. S., Galetti, M., Ceballos, G., Isaac, N. J. B., Collen, B. (2014). Defaunation in the Anthropocene. *Science,* 345(6195), 401–406. doi:10.1126/science.1251817.

Dixon, L., Kavanagh, R., & Kroopf, S. (Producers), Burger, N. (Director). (2011). *Limitless* (Motion picture). United States: Virgin Produced/Rogue.

Fagan, B. (2012). *Cro-Magnon: How the Ice Age Gave Birth to the First Modern Humans*. New York: Bloomsbury Press.

Gaudin, Sharon (2014). Stephen Hawking says AI could 'end human race'. *Computerworld*, Dec. 13, 2014; http://www.computerworld.com/article/2854997/stephen-hawking-says-ai-could-end-human-race.html.

Hall, Charles A. S. (2011). Synthesis to special issue on new studies in EROI (Energy Return on Investment). *Sustainability,* 3, 2496–99; doi:10.3390/su3122496.

Hansen, J., Sato, S., Russell, G., & Kharecha, P. (2013a). Climate sensitivity, sea level, and atmospheric carbon dioxide. *Phil. Trans. R. Soc. A* 2013, 371, 20120294. doi:10.1098/rsta.2012.0294.

Hansen, J., Karecha, P., Sato, M., Masson-Delmotte, V., Ackerman, F., Beerling, D. J., ... Zachos, J. C. (2013b). Assessing ''dangerous climate change'': Required reduction of carbon emissions to protect young people, future generations and nature. *PLOS One*, 8(12), e81648. doi:10.1371/journal.pone.0081648 .

Homer-Dixon, T. (2007). *The Upside of Down: Catastrophe, Creativity, and the Renewal of Civilization*. Toronto: Vintage/Random House.

Hughes, D. (2014). *Drilling Deeper*. Santa Rose, CA: Post Carbon Institute. http://www.postcarbon.org/publications/drillingdeeper/.

Inman, M. (2013). The true cost of fossil fuels. *Scientific American*, April, 58–61.

Inman, M. (2014). Natural Gas: The Fracking Fallacy. *Nature* 516(7529), 4 December, 28–30. doi:10.1038/516028a.

Intergovernmental Panel on Climate Change (IPCC). (2014). Synthesis Report: Summary for policy makers. http://www.ipcc.ch/.

Jacobson, M. Z. & Delucchi, M. L. (2011). Providing all global energy with wind, water, and solar power, Part I: Technologies, energy resources, quantities and areas of infrastructure, and materials. *Energy Policy,* 39, 1154–1169; doi:10.1016/j.enpol.2010.11.040.

Kaufman, L. (1986). Why the ark is sinking. In L. Kaufman & K. Mallory, K. (Eds.), *The Last Extinction* (pp. 1–41). Cambridge, MA: The MIT Press.

Kirshner, R. P. (2002). The Extravagant Universe: Exploding Stars, Dark Energy, and the Accelerating Cosmos. Princeton & Oxford: Princeton University Press.

Kolbert, E. (2014). *The Sixth Extinction: An Unnatural History*. New York: Picador.

Kuhn, T. S. (1970). *The Structure of Scientific Revolutions* (2nd ed.). Chicago, IL: University of Chicago Press.

Kurzweil, R. (2005). *The Singularity is Near: When Humans Transcend Biology*. London: Penguin Books.

Landauer, R. (1991). Information is physical. *Physics Today,* 44(5), 23–29.

McGlade, C. & Ekins, P. (2015). The geographical distribution of fossil fuels unused when limiting global warming to 2 °C. *Nature* 517(7533): 187–190. 2015 doi:10.1038/nature14016.

Motasharrei, S., Rivas, J., & Kalnay, E. (2014, April 2). Human and nature dynamics (HANDY): Modelling inequality and use of resources in the collapse or sustainability of societies. *Ecological Economics* 101, 90–102. doi:10.1016/j.ecolecon.2014.02.014.

Murphy, D. J., & Hall, C. A. S. (2010). Year in review—EROI or energy return on (energy) invested. *Ann. N.Y. Acad. Sci.,* 1185, 102–118. doi:10.1111/j.1749-6632.2009.05282.x.

Peacock, K. A. (1999). Staying out of the lifeboat: Sustainability, culture, and the thermodynamics of symbiosis. *Ecosystem Health,* 5(2), 91–103.

Peacock, K. A. (2008). *The Quantum Revolution: A Historical Perspective*. Westport, CT: Greenwood Press.

Peacock, K. A. (2011). Symbiosis in ecology and evolution. In K. deLaplante, B. Brown, & K. A. Peacock (Eds.), *Philosophy of Ecology* (218–250). Amsterdam: Elsevier.

Pollard, D., DeConto, R. M., & Alley, R. B. (2015). Potential Antarctic Ice Sheet retreat driven by hydrofracturing and ice cliff failure. *Earth and Planetary Science Letters,* 412, 112–121; doi:10.1016/j.epsl.2014.12.035.

Schneider, E. D., & Sagan, D. (2005). *Into the Cool: Energy Flow, Thermodynamics, and Life*. Chicago & London: University of Chicago Press.

Smolin, L. (2006). *The Trouble With Physics: The Rise of String Theory, the Fall of a Science, and What Comes Next*. New York: Houghton Mifflin.

Tainter, J. A. (1988). *The Collapse of Complex Societies*. Cambridge, UK: Cambridge University Press.

Ward, P. D. (2007). *Under a Green Sky: Global Warming, The Mass Extinctions of the Past, and What They Can Tell Us About Our Future*. New York: Collins/Smithsonian Books.

Whatmough, G. A. (1996). The artifactual ecology: An ecological necessity. In K. A. Peacock, (Ed.), *Living With the Earth: An Introduction to Environmental Philosophy* (417–420). Toronto: Harcourt Brace & Co., Canada.

Wilson, E. O. (1992). *The Diversity of Life*. New York & London: W. W. Norton.

Woit, P. (2006). *Not Even Wrong: The Failure of String Theory and the Search for Unity in Physical Law*. New York: Basic Book.

Chapter 9
Computer Simulations as a Technological Singularity in the Empirical Sciences

Juan M. Durán

9.1 Introduction

To talk of a 'singularity' evokes the idea of a natural occurrence beyond which the course of humanity has changed in significant ways. The Big Bang is one good example of such a singularity that affects our future on physical and existential levels. Now, to talk of a 'technological singularity' seems to evoke something else, something related to an occurrence that is the byproduct of human activity. Philosophers have not yet come to an agreement on the notion of a technological singularity. What seems to be common ground, however, is the idea that in order to speak of a technological singularity, humans must somehow be displaced from the center of the production of knowledge. Much of current literature asserts that the origin of such a displacement is related to the introduction of computers and their pervasive use in our modern life. Either by means of the emergence and evolution of superintelligent machines (Good 1965; Vinge 1993), or by means of powerful computers that enhance human cognitive capacities (Kurzweil 2012), computers are changing the way we gather and organize information about our world. Thus understood, the notion of technological singularity is intended to underscore the existence of an epistemological barrier beyond which humans cannot trespass, and where computers become the center of the epistemological endeavor.[1]

The specialized literature has mainly focused on problems of singularity stemming from the philosophy of mind (Turing 1950), philosophy of economics (James Miller 2012), and ethics (Hans Jonas 2011), just to mention a few. This paper, instead, centers on the junction between the philosophy of science and the still young but vigorous philosophy of computer simulations. Concretely, I am interested in focusing

[1]Since the context is clear, from now on I will use the terms 'technological singularity' and 'singularity' interchangeably.

J.M. Durán (✉)
HLRS - University of Stuttgart, Stuttgart, Germany
e-mail: duran@hlrs.de

V. Callaghan et al. (eds.), *The Technological Singularity*,
The Frontiers Collection, DOI 10.1007/978-3-662-54033-6_9

the problem of technological singularity from the viewpoint of computer simulations and their pervasive use in the empirical sciences. The general question that is being posed here: under what conditions can computer simulations be considered a technological singularity in the empirical sciences?

In this context, we should first note that I am assuming that at least some computer simulations do qualify as a technological singularity. This fact is, to my mind, uncontroversial. It should not take much effort to find a case in current scientific practice where a computer simulation advanced the general scientific knowledge of a target system. However, to assume that some computer simulations qualify as a technological singularity presupposes an understanding of what a technological singularity entails. Allow me, then, to begin by drawing a minimal working characterization of what makes a computer simulation qualify as a technological singularity.

> A computer simulation is a technological singularity if humans have been displaced from the center of production of knowledge, and if such knowledge is reliable to the extent of being usable without further sanctioning.

Although this characterization needs clarification, it provides the first basic intuitions on the issue. To begin with, displacing humans from the production of knowledge is necessary, although not sufficient, for considering computer simulations a singularity, as is demonstrated by the use of computer simulations for 'face recognition.' Although such a simulation displaces humans in some clear respects, such as in simulating how the human brain would analyze the large amount of data required for face recognition, it does not render better information (e.g., the computer simulation could not recognize a human face as fast and as accurate as a human could). Conversely, a computer simulation could provide knowledge of an empirical target system without necessitating the displacement of humans from the center of production of knowledge. A simple example of this is a simulation of an orbiting satellite using Newtonian mechanics. Humans have been calculating these orbits for many years and there are no good reasons for thinking that a simulation necessarily displaces humans from such simple calculations. As for demanding a reliable production of knowledge, it is straightforward that such a claim is needed for entrenching the idea of singularity. Although still in rough form, these intuitions concede the existence of conditions under which computer simulations qualify as a technological singularity. Could we evaluate more precisely what these conditions are?

In 2009 Paul Humphreys wrote a cogent paper defending the novelty of computer simulations in the philosophical arena. In that work, he introduced the notion of 'anthropocentric predicament' as the question about "how we, as humans, can understand and evaluate computationally based scientific methods that transcend our own abilities" (Humphreys 2009, 617). At first glance, the anthropocentric predicament aims at fathoming the methods used by scientists for the evaluation of computer simulations. Under this interpretation, the anthropocentric predicament is tailored to human cognitive capacities, and therefore to an anthropocentric epistemology. But Humphreys' intentions are precisely the opposite. To his mind, computer simulations come to take the place of humans in the production of knowledge and, as such, "an exclusively anthropocentric epistemology is no longer appropriate because there now exist superior, non-human, epistemic authorities" (Humphreys 2009, 617).

Understood in this second sense, the anthropocentric predicament is the route to considerations about computer simulations as a technological singularity in the empirical sciences. However, as presented, it is silent on the conditions under which a simulation produces knowledge of an empirical target system that is as reliable as a human. This paper, then, draws from the anthropocentric predicament and addresses what I believe are the basic conditions for considering computer simulations as a technological singularity. I begin by showing how, under the correct epistemological and methodological conditions, computer simulations are reliable producers of simulation results about the empirical world. For this, I propose to revisit the anthropocentric predicament as elaborated by Humphreys and argue, following the author, that an exclusively anthropocentric epistemology is no longer appropriate in a context where reliable knowledge is also delivered by computer simulations. This first part of the paper shows in what sense the anthropocentric predicament is key for understanding computer simulations as a technological singularity. A second and more extended part assimilates the previous outcomes and deals with the problem of reliability at face value. For this, I rely on Alvin Goldman's work on *process reliabilism*. The core idea is that under the right epistemological and methodological conditions, computer simulations are reliable processes that produce knowledge about the empirical target system. The challenge in this second part is to show what those conditions are and why we are justified in believing in the reliability of computer simulations. I then propose to approach this issue by firstly adopting a representationalist viewpoint, where a computer simulation is a reliable process if it correctly represents the target system (and, as I will also argue, correctly computes the simulation model). Thus understood, this approach greatly reduces the number of computer simulations capable of qualifying as a technological singularity. This outcome, although correct, seems to be at odds with some uses of computer simulations in daily scientific practice. In effect, there are concrete cases where computer simulations do provide reliable knowledge, even when the implemented simulation model does not fully represent the empirical target system. The conceptual shift, then, is to consider the many 'epistemic functions' that computer simulations perform and allow them to qualify as a technological singularity. Good examples of such epistemic functions are prediction and exploratory strategies. Due to lack of space, however, I will leave unanswered the question of whether a non-representationalist computer simulation qualifies (and, if so, under what conditions) as a technological singularity in the empirical sciences.

9.2 The Anthropocentric Predicament

In 2009, Roman Frigg and Julian Reiss warned us of the growth of overemphasized and generally unwarranted claims about the philosophical importance of computer simulations. This growth was reflected in the increasing number of philosophers convinced that the philosophy of science, nourished by computer simulations, required an entirely new epistemology, a revised ontology, and novel semantics. Although the authors admit the importance that computer simulations have in contemporary theoretical and practical science, they believe that computer simulations hardly call into question the basic philosophical principles of understanding science and conceptualizing reality (Frigg and Reiss 2009, 594–595). I share with Frigg and Reiss the puzzlement on this issue. It is hard to sustain that a new scientific method (instrument, mechanism, etc.), however powerful and novel it might be, could all by itself imperil the current state of philosophy of science. It is still an open question, however, whether or not Frigg and Reiss have correctly interpreted the underlying claims of the authors they criticize.

An important consequence of the article was the reactions that arose within the philosophical community. Paul Humphreys, for instance, directly engaged Frigg and Reiss with an answer. He claimed that Frigg and Reiss' assertions obscured the challenges that computer simulations pose for the philosophy of science. The core of Humphreys' reply was to recognize that the question about the novelty of computer simulations has two sides: one side which focuses on how traditional philosophy illuminates the philosophical study of computer simulations (e.g., through a philosophy of models and a philosophy of experiment, as Frigg and Reiss claim); and another side which exclusively focuses on aspects of computer simulations in and of themselves, that is, philosophical questions stemming from the very object of study regardless of any clarification from a more familiar philosophy. It is this second way of looking at the issue that gives philosophical importance to computer simulations.

Humphreys, then, elaborates on a set of novelties ascribed to computer simulations that constitute distinctively new methods in the scientific and philosophical arena. This set includes, among others, the 'epistemic opacity' of computer simulations relative to a cognitive agent, that is, the impossibility of knowing all the epistemically relevant constituents acting during a simulation. In short, being epistemically opaque means that, due to the complexity and speed of the computational process, no cognitive agent could follow the entire simulation. A second novelty that chimes with epistemic opacity is the 'temporal dynamics' of computer simulations. This concept has two possible interpretations. Either it refers to the necessary computer-time to solve the simulation model, or it stands for the temporal development of the target system as represented in the simulation model. A good example of both interpretations of temporal dynamics is a simulation implementing a model of the dynamics of the atmosphere that takes, say, 100 days to compute.

These two novelties nicely illustrate what is typical of computer simulations, namely, their inherent complexity, as the case of epistemic opacity and the first interpretation of temporal dynamics reveals; and the inherent complexity of the target systems, as the case of the second interpretation of temporal dynamics shows. Now, what is common between these two novelties is, in turn, that they both entrench computers as the main epistemic authority by displacing humans from the center of the production of knowledge. Indeed, either because the process of computing is too complex for us to follow or because the target system is too complex for us to comprehend, computers become the exclusive source for gathering information about the world.

Humphreys called this feature the *anthropocentric predicament*, which refers to the idea of understanding the world from a non-human perspective by representational intermediaries tailored to human cognitive capacities (Humphreys 2009, 617). The anthropocentric predicament, then, gets its support from the view that scientific practice only progresses because new methods are available for handling large amounts of information. In former times, the amount of information collected by a given discipline was, to a certain extent, manageable for the scientists. The astronomers could keep their books and stellar maps, and even perform many different kinds of calculations by controlling every step of the process. Today's scientific practice, however, handles enormous amount of information that are virtually impossible to manage without the aid of technology. A good example of this is the Diffuse Infrared Background Experiment on the NASA Cosmic Background Explorer satellite which produces 160 million measurements per twenty-six week period (Wright 1992, 231). The new technology which Humphreys has in mind is, surely, the digital computer. An equally important aspect of the anthropocentric predicament is that it requires computer simulations to represent an empirical target system. Such a requirement seems to be a natural consequence of the issue at stake, for displacing humans from the center of production of knowledge requires that knowledge to be grounded in the structure of the target system.

Thus understood, the anthropocentric predicament is the route to entrenching computer simulations as a technological singularity in the empirical sciences. Nevertheless, in order to be a singularity, it is not enough for computer simulations to simply set humans apart from the epistemic enterprise, it also requires the simulation results to be epistemically on *a par* with the data that could have been produced by a human agent. Indeed, a computer simulation that represents a sophisticated Ptolemaic model could displace humans from the center of knowledge insofar as it produces accurate results that, other things being equal, humans could not produce by themselves. However, such a computer simulation could not be counted as a reliable source of information about the planetary movement, and therefore it could not seriously be considered as a reliable producer of knowledge. Computer simulations are not epistemic leverage that turns any implemented model into insight about the empirical world, nor can they simply displace human agents from the center of production of knowledge by representing inherently complex target systems and readily producing results of intricate models. In order to be a singularity, computer simulations must produce results that effectively lead to knowledge about their empirical target systems. To this end, computer simulations must be conceptualized as reliable processes in a specific sense yet to be determined. It is in this precise sense that the anthropocentric predicament fails to ground computer simulations as a singularity, remaining silent on the conditions under which they are reliable producers of knowledge about the empirical world.

The following section addresses the conditions for a reliable computer simulation. To my mind, a reliable computer simulation is one that renders, most of the time, valid simulation results about the empirical target system. A reliable computer simulation, then, provides the justification that our beliefs about the simulation results are, most of the time, true rather than false of that empirical system, leading to knowledge about such a target system. The question about singularity, then, is subsumed into the question of what it means to have a reliable computer simulation.

Let it be noted that singularity entails reliability, although the converse is not necessarily true. Many computer simulations are reliable in the sense just given, although they do not qualify as a singularity. The reason for this is fairly simple. A computer simulation can be reliable even when it is sanctioned *after* the computation of the simulation results. For instance, (Jenkins et al. 2014) present a simulation of self-assembling DNA-coated spheres. The unusual feature of this simulation is that it has thousands of configurations that are as energetically favorable as the real experiment. Only expert knowledge can determine which simulation is more accurate than others. It follows that the simulation results are sanctioned *after* the computation of the simulation, and therefore the simulation cannot be considered a singularity. In the face of this, let me call the juncture of all methods that collectively grant reliability to the simulation *before* its execution on the computer, and therefore *before* obtaining the simulation results, the *pre-computed reliability stage*. There is also a *post-computed reliability stage*, which refers to methods for sanctioning the simulation results *after* they have been obtained and, therefore, requiring the intervention of an agent. Since singularity is only concerned with the pre-computed reliability stage of a computer simulation, there is no need to address issues related to post-computed reliable stages. Allow me now to discuss in more detail what entails a pre-computed reliability stage.

9.3 The Reliability of Computer Simulations

Epistemologists have taught us that questions about knowledge have their roots in the notion of truth and epistemic justification. While it is widely agreed that what is false cannot be known, there is less consent on what it means to say that we know something and why we are justified in believing so. In this paper I am only interested in analyzing the justification of believing that the simulation results are valid of a target system. To be a *valid simulation result* is to match, with more or less accuracy, the (theoretically) measured and observed values of the empirical target system.

With these ideas in mind, allow me to address the philosophical importance of epistemic justification for the singularity hypothesis. Let me begin by asking in what specific sense are we justified in believing that a computer simulation renders knowledge about the empirical world? A suitable answer is to consider the computer simulation as a belief-forming process, that is, as a process capable of producing results that are, most of the time, valid of the empirical target system. If such results can be produced, then we can say that we are justified in believing the simulation results and, as such, in claiming for knowledge of the empirical target system. In plain words, if the simulation results are acceptably close or similar to real-world measurements, observations, or even pen-and-paper calculations, then we are entitled to claim empirical knowledge of that target system.

In epistemology, such an account is known as *process reliabilism*, where one of the major contributors and theorist has been Alvin Goldman. In its simplest form, reliabilism says that a belief is justified in the case that it is produced by a reliable process, where 'reliable' here means a process that produces, most of the time, truths (Goldman 1979). For instance, we know that '$2 + 2 = 4$' because the reasoning process involved in addition is, under normal circumstances and within a limited set of operations, a reliable process. According to Goldman, then, there is nothing accidental in knowledge that is produced by a reliable process. Reinterpreting process reliabilism for computer simulations, we can say that a simulation is a reliable process if it produces results that are, most of the time, valid of the target system. Following Goldman, there is nothing accidental about believing that a computer simulation produces valid results of a target system (provided that certain conditions are fulfilled) and, consequently, about the claim that we obtain knowledge of that target system. Thus understood, process reliability is resolved as the belief-forming process that renders simulation results as valid for the intended empirical target system. This point can be easily illustrated by considering CS_R as the simulation results and RW_D as the data of its target system. According to reliabilism, then, we are justified in believing the results of the computer simulation if $|CS_R - RW_R| \cong 0$, that is, if the simulation results approximate the real data measured, observed, or computed. To put the same ideas in a rather different form, the reliability of computer simulations is built on the quantitatively assessed accuracy of its results. Such accuracy is obtained by two sources, namely, the representational capacity of the simulation and a relatively error-free computation. The question about the reliability of computer simulations and, therefore, of knowledge is now shifted to understanding these two sources.

Let me motivate this issue by briefly enumerating which computer simulations could not be regarded as reliable.

1. Cases of misrepresentations:
    a. Any computer simulation that implements a known false model (e.g., the Ptolemaic model of the solar system) could not be expected to render knowledge of planetary movement.
    b. Any computer simulation that has no representational underpinning of the target system, such as heuristic simulations. (e.g. the Oregonator is a simulation for exploring the limits of the Belousov chemical reaction. Such a simulation implements a model whose system of equations is *stiff* and therefore it might lead to qualitatively erroneous results (Field and Noyes 1974, 1880)). A particular case of this is:
    i. Any computer simulation that renders *unrealistic simulated results*, that is, results that cannot represent an empirical target system (e.g., a computer simulation implementing a Newtonian model setting the gravitational force

to $G = 1\,\mathrm{m}^3\mathrm{kg}^{-1}\mathrm{s}^{-2}$).[2] Such simulations violate the laws of nature and cannot be considered empirically accurate.

2. Cases of miscalculations:
   a. A computer simulation that miscalculates due to large round-off errors, large truncation errors, and other kinds of artifacts in the calculation, such as ill-programmed algebraic modules and libraries. Such software errors cannot stand for valid simulation results of the target system.
   b. A computer that miscalculates due to physical errors, such as an ill-programmed computer module or a malfunctioning hardware component. Similar to (2.a) above, these types of errors warrant invalid simulation results and, as such, do not render knowledge of the target system.

A generally valid principle in computer simulations is that there are no limits to the imagination of the scientists. This is precisely the reason why simulations are, one might argue, facilitating the shift from a traditional empirically-based scientific practice into a more rationally-based one. However, neither of the examples described above fit the conditions for a pre-computed reliable simulation. While simulations belonging to case (1.a) are insufficient for an accurate representation the empirical target system, those belonging to case (1.b) are highly contentious. The latter case, as is illustrated by the Oregonator example, is trusted only insofar as the results are subject to the subsequent acceptance by experts. Whenever this is the case, the simulation automatically fails to classify as pre-computed reliable and, therefore, violates the basic assumption of the singularity hypothesis. Although much of current scientific practice depends on these kinds of simulations, they do not comply with the minimal conditions for being a singularity, and therefore they inevitably fail to qualify as one.

In the same manner, heuristic simulations must not necessarily render invalid results. For this reason, they are useful simulations for exploring the mathematical limits of the simulation model, as well as the consequences of an unrealistically constructed law of nature, among other uses. Such simulations facilitate the representation of counterfactual worlds, thought experiments, or simply fulfill propaedeutic purposes, but given our current conception of technological singularity, they do not qualify as such.

Besides their representational features, computer simulations are also part of the laboratory *instrumentarium*, and as such are inevitably exposed to miscalculations of different sorts. A first classification of errors divides them into *random errors*, such as a voltage dip during computation, or a careless laboratory member tripping over the power cord, and *systematic errors*, that is, errors that are inherently part of the simulation. Random errors have little philosophical value. Their low probability

[2]Unrealistic results are not equivalent to erroneous results. In this case, the results are correct of the simulation model although they do not represent any known empirical system.

of occurrence, however, do make a small contribution (although negligible) to the frequency of a process of producing beliefs that are false rather than true. Systematic errors, on the other hand, can be subdivided into *logical errors* (i.e., errors in the programming of software, such as the errors illustrated in (2.a) above), and *hardware errors* (i.e., errors related to the malfunctioning of the physical component of the computer, exemplified in (2.b) above).

This paper concedes that miscalculations do occur in the practice of computer simulations, but assumes that they are rare and rather negligible for the overall evaluation of the reliability of computer simulations. The reason is that over the years of technological advancement, computers have become less prone to suffer from failure. A host of recovering procedures such as duplication and redundancy mechanisms for critical components, functions, and data grant this dependability. Also, new techniques in the design and practice of programming, as well as the plethora of programming languages and expert knowledge at the programmers' disposal facilitate the assertion that computers are relatively error-free and fail-safe instruments. As a working assumption, then, I take that computer simulations are stable instruments that, most of the time, do not incur in calculation errors that might alter its results (or, if they do, such errors are entirely negligible).

9.3.1 Verification and Validation Methods

The bluntly false and highly speculative examples used above constitute only a small portion of computer simulations used in scientific practice. For the most part, scientists interpret, design, and program the simulation of an intended target systems with remarkable representational accuracy and on stable instruments. This paper concedes this much. We must carefully distinguish, however, simulation results that require further epistemic sanctioning from results whose validity has been granted during a pre-computed reliability stage. An argument is advanced to the effect of addressing *verification* and *validation methods* applied during the pre-computed reliability stage that grant validity to simulation results.

Verification and validation methods are at the basis of claims about the reliability of computer simulations. They build on the confidence and credibility of simulation results, and in this respect, understanding their uses and limits is central for claims about singularity. While verification methods substantiate the scientist's belief that the mathematical model is correctly implemented and solved by the simulation, validation methods provide evidence that the simulation results match, with more or less accuracy, empirical data. Let us take a closer look at what comprises each method.

The American Society of Mechanical Engineers (ASME), along with other institutions, adopted the following definition of verification: "[t]he process of determining that a computational model accurately represents the underlying mathematical model and its solution" (ASME 2006, 7). Thus understood, verification could be obtained in two ways: by finding evidence that the algorithms are working correctly, and by measuring that the discrete solution of the mathematical model is accurate.

The former method is called *code verification*, while the latter is known as *calculation verification*. The purpose of making these distinctions is to categorize the set of methods for the assessment of correctness of the computational model with respect to the mathematical model, as opposed to assessing the adequacy of the mathematical model with respect to the empirical system of interest. Code verification, then, seeks to remove programming and logic errors in the computer program and, as such, it belongs to the design stages of the computational model. Calculation verification, on the other hand, seeks to determine the numerical errors due to discretization approximations, round-off errors, discontinuities, and the like. Both code verification and calculation verification, are guided by formal and deductive principles, as well as by empirical methods and intuitive practice (or by a combination of both).

Validation, on the other hand, has been defined as "[t]he process of determining the degree to which a model is an accurate representation of the real world from the perspective of the intended uses of the model" (ASME 2006, 7). Validation, then, is somehow closer to the empirical system since it is concerned with the accuracy in the representation of such a model. Validation methods also make use of benchmarks or reference values that help establish the accuracy of the simulation results. Benchmarking is a technique used by computer scientists for measuring the performance of a computer system based on comparisons between simulation results and experimental data. The simplest way to obtain such data comes from performing traditional empirical experiments. Now, since the true value of the empirical target system cannot always be absolutely determined, it is an accepted practice to use a reference value obtained by traditional measurements and observational procedures. A different situation is when the value of the target system can be theoretically determined, as is the case in quantum mechanics where the value of the position of atoms are obtained by theoretical methods. In such cases, the simulation results can be validated with high accuracy. In the same vein, results from different but related simulations could also be used for validation purposes, as these results can be easily compared to the simulation results of interest. Validation, then, aims at providing proof of the accuracy of simulation results with respect to the empirical target system of interest. Thus understood, the appeal to validation methods as grounds for reliable processes brings out questions that lurk behind inductive processes. The general concern is that such methods only allow validation up to a certain number of results, that is, up to those of which we have previous data. Due to their comparative nature, these methods do not provide mechanisms for validating new, unknown results. It follows that validation is a method for assisting in the detection of errors, but not designed for detecting misrepresentations of the target system.

It should not be expected, however, that during an actual verification or validation process scientists decouple these methods. The multiple problems related to mathematical representation, mathematical correctness, algorithm correctness, and software implementation make the entire enterprise of verification and validation highly interwoven processes. Moreover, code verification by formal means is virtually impossible in complex and elaborated simulations. Let it also be noted that not all verification and validation methods are performed at the same stages of design and output of a simulation. Some verifications are only carried out during design stages, while others, such as *manufactured solutions*, depend on the intervention of an agent. Manufactured solutions are custom-designed verification methods for highly accurate numerical solutions to partial differential equations (PDEs). It consists in testing numerical algorithms and computer codes by finding solution functions that have altered the implemented PDEs, but which also satisfy such equations. As William Oberkampf and Timothy Trucano indicate, "[a manufactured solution] verifies many numerical aspects in the code, such as the mathematical correctness of the numerical algorithms, the spatial-transformation for the grid generation, the grid-spacing technique, and the absence of coding errors in the software implementation" (Oberkampf and Trucano 2008, 723).

In a similar fashion, validation methods might focus on the design stages as well as on the output of a simulation. It is easier to devise validation methods requiring the intervention of an agent, for the construction and subsequent use of benchmarks requires such involvement. Examples of this abound in the literature and there is no need to discuss this point any further. One could always consult Oberkampf and Trucano's list for the documentation of benchmarks, all of which must be in place for successfully warranting accuracy of the computed results (Oberkampf and Trucano 2008, 728).

9.4 Final Words

In this paper I defended the idea that, under the right conditions, computer simulations are reliable processes that produce, most of the time, valid simulation results. Valid simulation results are taken as knowledge of the target system, facilitating the claim that computer simulations are a technological singularity. Now, in order to fully qualify as a technological singularity, such results must not be sanctioned by a human agent. On the face of it, the number of simulations that qualify as a technological singularity has been reduced to a few well established cases with representational underpinning and error-free computations. Such representation and error-free computations are grounded on verification and validations methods, as elaborated above.

One immediate consequence is that the universe of computer simulations has been significantly reduced. At first, this outcome might strike one as an undesirable and anti-intuitive consequence. One might think that many computer simulations are being used today as reliable processes producing valid results of a given empirical system, and that there is no special problem in doing so. The general trend nowadays is to overthrow humans as the ultimate epistemic authority, replacing them with computer simulations (Sotala and Yampolskiy 2015). A good example of this is the simulation of the spread of influenza, as expounded by Ajelli and Merler (2008), where two different kinds of computer simulations provide knowledge of a hypothetical scenario. In some situations, in effect, no further sanctioning is needed and the information provided by these simulations is used as obtained. However, contrary to appearances, the vast majority of cases get their results sanctioned after they have been produced, undermining the possibilities of becoming a technological singularity. One might safely conclude that the number of computer simulations that qualify as a singularity are, indeed, limited.

Admittedly, much more needs to be said in both directions, namely, what grounds computer simulations as a technological singularity (especially regarding verification and validation methods) as well as how current scientific practice accommodates this philosophical view. Equally important is to elaborate on cases such as Ajelli et al., where the simulation seems to be a singularity if used in certain situations and fails to be one in some others.

Acknowledgements This article was possible thanks to a grant from CONICET (Argentina). Special thanks also go to Pío García, Marisa Velasco, Julián Reynoso, Xavier Huvelle, and Andrés Ilcic (Universidad Nacional de Córdoba - Argentina) for their time and comments.

References

Marco Ajelli and Stefano Merler. The impact of the unstructured contacts component in influenza pandemic modeling. *PLoS ONE*, 3(1):1–10, 2008.

ASME. Guide for verification and validation in computational solid mechanics. Technical report, The American Society of Mechanical Engineers, ASME Standard V&V 10-2006, 2006.

R. J. Field and R. M. Noyes. Oscillations in chemical systems IV: Limit cycle behavior in a model of a chemical reaction. *Journal of Chemical Physics*, 60:1877–1884, 1974.

Roman Frigg and Julian Reiss. The philosophy of simulation: Hot new issues or same old stew? *Synthese*, 169(3):593–613, 2009.

Alvin I. Goldman. What is justified belief? In G. S. Pappas, editor, *Justification and Knowledge*, pages 1–23. Reidel Publishers Company, 1979.

I. J. Good. Speculations concerning the first ultraintelligent machine. In F. Alt and M. Rubinoff, editors, *Advances in Computers*, volume 6. Academic Press, 1965.

Paul W. Humphreys. The philosophical novelty of computer simulation methods. *Synthese*, 169(3):615–626, 2009.

Ian C. Jenkins, Marie T. Casey, James T. McGinley, John C. Crocker, and Talid Sinno. Hydrodynamics selects the pathway for displacive transformations in DNA-linked colloidal crystallites. *Proceedings of the National Academy of Sciences of the United States of America*, 111:4803–4808, 2014.

Hans Jonas. Technology and responsibility: Reflections on the new task of ethics. In Morton Winston and Ralph Edelbach, editors, *Society, Ethics, and Technology*, pages 121–132. Cengage Learning, 2011.

R. Kurzweil. *The singularity is near: When humans transcend biology*. New York: Viking, 2005.

James Miller. Some economic incentives facing a business that might bring about a technological singularity. In Amnon H. Eden, James H. Moor, Johnny H. Soraker, and Eric Steinhart, editors, *Singularity Hypotheses: A Scientific and Philosophical Assessment*. Springer, 2012.

W. L. Oberkampf and T. G. Trucano. Verification and validation benchmarks. *Nuclear Engineering and Design*, 238(3):716–743, 2008.

A. M. Turing. Computing machinery and intelligence. *Mind*, 59(236):433–460, 1950.

V. Vinge. The coming technological singularity: How to survive in the post-human era. In *Proc. Vision 21: interdisciplinary science and engineering in the era of cyberspace*, pages 11–22. NASA: Lewis Research Center, 1993.

E. L. Wright. Preliminary results from the FIRAS and DIRBE experiments on COBE. In M. Signore and C. Dupraz, editors, *The Infrared and Submillimeter Sky After COBE. Proceedings of the NATO Advanced Study Institute. Les Houches, France, Mar. 20–30, 1991*, pages 231–248. Kluwer, 1992.

Kaj. Sotala and Roman. V. Yampolskiy. Responses to catastrophic AGI risk: a survey *Physica Scripta*, 90(1), 2015.

Chapter 10
Can the Singularity Be Patented? (And Other IP Conundrums for Converging Technologies)

David Koepsell

10.1 Introduction

Many of the societal controls and policy considerations discussed in relation to the development of "The Singularity" or artificial general intelligence (AGI) relate to potential harms and risks. There is one important area, however, where *current* legislation and policies pose a risk to its development: the laws of intellectual property. Intellectual property (IP) is now taken for granted as somehow necessary for innovation, but it poses a considerable threat to both the eventual development of the Singularity, and to our rights, duties, and obligations once such a Singularity emerges. If we are to properly anticipate and guide the development of AGI in positive ways, we must come to terms with the role that IP plays in either promoting or hindering innovation, the impact it will have on AGIs as well as their developers, and manners in which might better manage IP to allay some of the anticipated risks.

Many people credit the development of the legal institution we call "intellectual property" with helping to propel us into the modern, technological age. Before intellectual property (IP) was invented, there was no means available for people to prevent others from utilizing their ideas, aside from force. This is because ideas cannot ordinarily be monopolized, except by secret-keeping, and even then they are prone to independent discovery or thought. Because of the inherent uncontainability of ideas, states have, within the past couple hundred years, devised legal monopolies over expressions in various forms under the belief that such monopolies would help encourage innovation and immigration of highly skilled and inventive people.

D. Koepsell (✉)
Research and Strategic Initiatives, COMISIÓN NACIONAL DE BIOÉTICA, Mexico city, Mexico
e-mail: drkoepsell@gmail.com
URL: http://davidkoepsell.com

D. Koepsell
Universidad Autonoma Metropolitan, Xochimilco, Mexico

V. Callaghan et al. (eds.), *The Technological Singularity*,
The Frontiers Collection, DOI 10.1007/978-3-662-54033-6_10

The first such monopolies were "letters patent" issued by sovereigns to those who devised new and useful arts. These letters patent entitled their holders to monopolize a market for a term of years. Over time, various states began to formalize these sorts of processes, and introduced also monopolies for artistic works, believing that these incentives would help to encourage technological and economic growth.

Modern IP includes copyrights, patents, and trademarks. Each works slightly differently, and covers specific types of expressions. Patents are monopolies granted by the state to inventors of new, useful, and nonobvious utilitarian objects. The monopoly term for patents runs 20 years, after which the invention lapses again into the public domain. The copyright term, which covers aesthetic expressions, lasts the lifetime of the author plus an additional 70 years. Trademarks cover trade names, and fall outside of this discussion. Our interests here focus on the effects of both patents and copyright on the emergence and promise of the singularity, as it has been hypothesized by Ray Kurzweil and others. Both copyrights and patents are pertinent due to some oddities in the law of IP that I began to explore nearly 20 years ago, and which remain unresolved. But more to the point, patents are the primary threat, at least in their current forms, to both the emergence of the singularity and its achieving its full promise.

10.2 A Singular Promise

The singularity can be understood in numerous ways, and people more technically inclined have devised excellent explications of its potential nature. For our purposes, let us examine it in its most general possible forms and then discuss how various legal institutions such as copyrights and patents may impact on its eventual achievement. The technological singularity will involve a quantum leap in our technologies in such a way that society itself will change. This could happen through the realization of true artificial intelligence, which, if it can be created, will in all likelihood be unbounded by the biological limitations imposed upon our own intelligences. The singularity may also be achieved by way of nanotechnology or some other radical new approach to our material world such that objects can be programmed: our whole physical environment could be alterable either by ourselves, or merged with our artificial intelligences. Finally, the singularity may involve some transhumanist future in which we ourselves become the objects of our technological change, capable of super-intelligence, strength, or longevity… or even immortality. The singularity may involve some combination of all of these, either to completion or partially realized. In any event, any or all of these technological achievements will likely usher in radical challenges to our present cultures and institutions. How we manage our current institutions, including laws, will likely have some impact too on the eventual realization of the singularity, and how it may then affect us once achieved, as we shall see.

Technological advances have necessarily wrought significant reflection, if not always rapid change, in the ways that we consider the roles of law in the field of innovation. Specifically, of late there has been some shifting of approaches to innovation and how best to encourage and protect investments. Open Source is now a serious option for even the biggest, most established companies seeking to create economic climates more suitable to commercial exploitation of some niches. This is perhaps because legal monopolies may discourage cooperation and competition, both of which may simultaneously be viable and profitable means of building a nascent technology's acceptance and success. Even early in the growth of the automobile industry, competing companies realized that legal monopolies could stifle the development of their technologies, and so entered into "patent pools" and other cooperative means to prevent "patent thickets" and encourage competition and the technological growth it can promote. The technological singularity, if it is underway, or if it is still far off on the horizon, will likely lead to similar legal, social, and institutional innovations if the current law of IP threatens to stifles its achievement, as I suggest below.

10.3 Intellectual Property

I have spent the better part of 15 years criticizing IP law in its current form for a variety of reasons, some pragmatic, and others essential. The law as it stands embodies a very confused metaphysics (by metaphysics, I mean the manner in which we classify its objects), and increasingly these metaphysical errors lead to absurd results. I will discuss my general issues with IP law briefly, and apply them to two particular issues I see as arising from IP's odd metaphysics and application specifically for the singularity hypothesis, namely: (1) does patent law as it now stands allow the patenting of artificial intelligences, and is this an ethical or practical problem, and (2) will IP protection apply to the creative products of artificial intelligences, and should it? These two complex but unresolved issues in the law of IP pose real potential impediments to its realization in the near future.

10.3.1 Some General IP Problems in Converging Technologies

I have written at some length about some of the practical and theoretical oddities and problems associated with IP law, some of which are particularly noticeable with the advent of digital and bio-technologies in the past few decades. I summed much of this up in my monograph, *Innovation and Nanotechnology: Converging Technologies and the End of Intellectual Property*. There I concluded that the train-wreck of IP, which began with the courts and patent offices granting simultaneous patents and copyrights (two, previously mutually-exclusive categories) to software, was culminating in the present innovative climate with a mandate for innovators to simply ignore and work around IP law entirely. This is because the courts and legislators who have drafted the current IP laws have interpreted them in ways that, while they might please IP lawyers and monopolists who have the means to acquire vast patent portfolios and thus venture capital (or worse, litigation awards), do not understand the underlying science or metaphysics of artifacts and nature. The result has been a slow usurpation of what many call the "scientific commons" and its gradual conversion into private property. This is especially problematic, I argue, for developing the technologies that will lead to the singularity.

Over time, the categories that originally bounded the objects of copyright and patent have been revealed to be flawed. Copyright was originally applied to "non-utilitarian" expressions, like works of art. Patents have been applied to "utilitarian" expressions, namely "*Whoever invents or discovers any new and useful process, machine, manufacture, or composition of matter, or any new and useful improvement thereof, may obtain a patent therefor, subject to the conditions and requirements of this title.*" These criteria have been interpreted over time to apply to nearly anything, including elements on the periodic table, business methods, software, algorithms, genes, and other seemingly problematic sorts of things have been granted patents that have been upheld. Even lifeforms may be patented, as long as they are not human, and as long as they have been "engineered" through some human intervention. Three explicit exclusions (other than the statutory exclusion of people) created by famous court cases and adopted by patent offices more or less around the world include: abstract ideas, products of nature, and natural phenomena. While these exclusions should have seemed more or less obvious under the original conceptions of patent law (which was to promote invention of new and useful things) the catch-all type of language in the US Patent Act's section 101 quoted above, and applied through various treaties to patents world-wide, have resulted in a blurring of the lines delineating nature from artifact, and courts have attempted through these exclusions to redraw those lines.

Unfortunately, the exclusions sought to be clarified have been further muddied through poor metaphysics in courts and patent offices. What counts as a "product of nature," for instance, is now more or less nothing. For instance, if you want to patent a product of nature, like the molecule O_2 (gaseous oxygen, created through such natural processes as photosynthesis) just come up with some man-made way to synthesize it. Thus, if I discover that I can concentrate O_2 through electrolysis of water, separating the O_2 from the H_2 in it, then under the current interpretation of the patent laws in the US and Europe, I can patent not just these new processes, but the products themselves. Suddenly, O_2 is an invention. All of which poses a real problem in nanotechnology, and converging technologies in general, because as the scales of manufacturing new and useful things shrink, the building blocks comprising new products as the singularity nears will be at the molecular level.

Consider the patent thickets that have emerged in the cell phone market, and now multiply by hundreds. This will slow down the convergence of technologies like nanotechnology and synthetic biology, and hinder the emergence of the singularity.

Indeed the law of IP seems to be reaching a crisis point as technologies converge upon the singularity. Written expressions and machines appear to be converging, blurring the line between aesthetic and utilitarian expressions in general. Biology and manufactures are converging, blurring the lines between artifacts and organisms. And the scales at which the building blocks of new artifacts are created are becoming smaller, making fundamental elements, deemed at one time to be products of nature, to be the medium for manufacturing nearly everything. Beyond these conundrums, however, lurk even more insidious ethical issues when (or should) the singularity emerge. What, for instance, will we do about patenting machines (or software) when they acquire sentience, and will we apply IP to the products of *their* imaginations?

10.3.2 Some Gaps in IP Relating to the Singularity

Before the twentieth century, there was no question about it, life forms could not be patented. This was largely because of the fact that technology had not yet advanced to the state of being able to consciously *create* new, non-obvious, and useful lifeforms. In 1930, the first US Plant Patent Act changed all that. Lobbied for by plant breeders as well as the likes of Thomas Edison himself, the Plant Patent Act provided the first monopoly protection under the Patent Act: 35 U.S.C. section 161: "Whoever invents or discovers and asexually reproduces any distinct and new variety of plant, including cultivated sports, mutants, hybrids, and newly found seedlings, other than a tuber propagated plant or a plant found in an uncultivated state, may obtain a patent therefor, subject to the conditions and requirements of this title." This was the first legislation to recognize and grant state-monopolies to living creatures. And why not? If the patent law states explicitly that any new, useful, and non-obvious composition of matter or manufacture should be able to be patented, then there's no particular reason embedded in these criteria to exclude lifeforms of any kind. Still, it wasn't until 1980 that the US Supreme Court extended this reasoning to non-plant life, in the seminal case of *Diamond v. Chakrabarty*.

Chakrabarty had "engineered" through selective breeding a bacterium that could digest petroleum, which would certainly be a useful new, non-obvious invention, helpful for cleaning up oil spills, etc. but the US Patent and Trademark Office refused to grant a patent for the organism citing no previous history, and thus implicit policy against patenting life. But the Supreme Court disagreed, and found the new organism patent-eligible, while stating that there were still some explicit exclusions to patent-eligibility, including: abstract ideas, products of nature, and natural phenomena. Interesting for the singularity hypothesis are the following: to what extent might the building-blocks of converging technologies fall under one of the *Chakrabarthy* exclusions, and should all engineered life-forms be patent-eligible?

One glaring gap in the law as it relates to converging technologies remains in the definitions of "abstract ideas," "products of nature," and "natural phenomena." Vagueness in the categories will continue to effect innovation in nanotechnology and artificial intelligence. The singularity, if it is technically achievable, will break down barriers among these categories, requiring us to reevaluate philosophically and practically what counts as abstract ideas, products of nature, and natural phenomena. Will artificial intelligences exhibit properties of nature, even if they are artificially produced. Will evolutionary forces, perhaps directing their future forms, require us to consider them to be under the guidance and direction of natural phenomena? If so, then do current IP exemptions mean that we cannot claim IP rights over future products? If life was a big exception under IP law prior to *Chakrabarthy*, and human life remains so now, what are the necessary and sufficient features of life that make it excludable?

We can deduce from the conflict between the statutory exclusion of human life from patentability, and the acceptance of other lifeforms as patentable after *Chakrabarthy*, that something special about humans prohibits owning them through IP. Of course, constitutional prohibitions against slavery should cover some forms of human ownership, preventing owning any one human, but extending this notion to IP conflates IP and ordinary property to an unfair extent. Simply put: IP does not confer ownership rights over tokens. That is to say, a patent holder doesn't have an ownership interest in any *instance* of his or her invention. Rather, they own the right to exclude others from profiting from the manufacture and first sale of the thing over which they hold IP rights. So IP ownership over engineered or partially-engineered humans ought not to be prohibited by the same rationale that prohibits slavery.

So what values would rationally prohibit ownership over types of some technical artefacts or processes and not others? One answer may lie in deontology and some Kantian concern over the effects of claims to ownership of types regarding sentient creatures as opposed to others. Are we entitled to *use* lifeforms as instrumentalities, or means to particular ends, or must we respect them (or at least some of them) as possessing certain inherent dignities which IP rights might offend? On its face, we clearly reject an expansive view of this thesis, since there are numerous lifeforms, even arguably "sentient" ones that we feel free to exploit as instrumentalities to varying degrees.

10.4 Limits to Ownership and Other Monopolies

Not everything is susceptible to just claims of monopoly or other control. Some limits to what may be owned are moral, and others are practical. Still others are legal creations, designed to achieve certain utilitarian ends. Most recently, the

Supreme Court in the U.S. has attempted to define better some existing limits to what could be claimed by patent. The *Abstract Idea*, *Natural Phenomena*, and *Laws of Nature* exceptions have been grappled with in the *Bilski*, *Mayo*, and *Myriad* cases between 2010 and 2014. First we will consider each of these exceptions and recent cases briefly, then examine their effect, if any, on singularity technology.

Intellectual property law was created in order to encourage and advance the march of the "useful arts." By granting temporary monopolies over the production and sale of new inventions and discoveries through various patent laws, governments attempt to promote economic development. IP law walks a delicate balance, however, between promoting invention and not impeding science. Scientific discovery typically focuses upon exploring and understanding the laws of nature, natural phenomena, and abstract ideas that govern the universe. This is perhaps a critical reason why the courts have carved out these subject matter exceptions to patent eligibility. Setting aside the moral problems posed by allowing monopolies over mere discoveries of naturally-occurring phenomena or laws, a practical and unwanted consequence might be to prevent researchers in basic sciences from conducting their research freely.

There have been certain realms in which such a monopoly was never deemed proper. The stated purpose of IP law is to promote invention, innovation, technological development and thus economic development. Science is typically the field viewed most appropriate for basic research into the working of nature, and the impetus for scientific discovery differs from that of commerce and industry. Science is impelled largely by scientific culture, which rewards discoveries about nature and its laws through academic careers, prizes, and often through the recognition and joy many of those who pursue scientific careers receive from uncovering nature's mysteries. Although competition and egos are also powerful motivators for scientists to pursue their work, for science to work correctly, a degree of openness and recognition of limits of knowledge and the contingency of present understanding, ensure that humility and cooperation remain part of the scientific culture as well.

By recognizing that abstract ideas, laws of nature, and natural phenomena cannot be patented, the US Supreme Court has upheld, despite the lack of explicit exclusion in the Patent Act, a boundary between science and technology; between what may be owned and what may not. The boundaries that have arisen in IP law will mean very little when it comes to the singularity, because the singularity promises to erase preexisting boundaries in revolutionary ways. Currently, one can patent life, just not humans. Currently, one can patent isolated and modified natural phenomena and products of nature, but not if they are, apparently, morphologically identical to natural products. What remains free for all are natural products qua natural products, and natural phenomena that have not been otherwise modified by man to serve some end. Where then does this leave the following possibility: converging technologies that result in artificially intelligent, self-replicating entities capable of modification through "evolutionary" means? Under current IP regimes, what repercussions will there be? What will be ownable through patent, and what not?

10.5 Owning the Singularity

Under the current IP regime described above, an artificially-created intelligent agent would be ownable, as long as it isn't somehow composed by cloning human DNA, which is so far illegal, and humans are specifically exempt from patent, even if engineered in some way. Such a product qualifies for patent as long as it meets the criteria of new, non-obvious, and useful. It is trivial that, given a lack of explicit exception for intelligent inventions, such an agent would be susceptible to IP claims. After all, who would invent such a useful thing without the patent incentive, one might argue. Be that as it may, patenting such an agent has significant moral repercussions and practical implications that we should consider.

Our hypothetical agent of the singularity will be able to be legally monopolized under patent by its creator. This means the following: no one may reproduce that agent without the patent holder's license. This may seem unproblematic at first; you made your intelligent agent through your own ingenuity and inventiveness, so why should anyone else be able to profit? But remaining questionable is whether that agent's ability (forgetting for the moment whether it has any rights) to reproduce is thwarted by the patent law. More complicated is the question of whether, if that agent can create new, non-obvious, and useful things of its own, that agent, the creator, or some other agent will have the ability to patent those new creations.

Whether we create some new, non-obvious, and useful agent through silicon-based artificial intelligence, uplifting other creatures, or through nano- or bio-tech, those agents will be ownable, to a degree. To the extent that they will themselves be impeded from reproducing duplicates of themselves, their rights as sentient being will be curtailed. They will be prevented from expressing one of the basic rights of sentient beings: self-reproduction. More curious will be the status of their inventions, if any. If they are sentient beings, and capable of creating new, non-obvious, and useful new things, will they be able to patent them, or will their own inventors? If the former, then curtailing their right to self-replicating reproduction seems unfair, whereas if the latter, then this seriously alters the manner in which we think inventors gain rights to the fruits of their intellectual labor.

A significant and perhaps insurmountable problem for IP in general will be the question of who shall rightly own the fruits of the inventiveness of inventive machines. It ought to cause us to question the foundations of IP law itself and its utilitarian intentions. A singularity will be able to compete with us humans in all realms, including invention and creativity. If so, and we don't bar non-humans from enjoying the benefits of a state-sponsored monopoly, singularity agents may well usurp our roles as inventors, and monopolize the realm of monopolies through the benefit of the state. This is a purely pragmatic concern, however. More theoretical is the ethical question regarding whether we could justly, though IP law, prevent artificial agents from reproducing themselves. Let's consider this first, and then re-raise the pragmatic concerns to suggest that IP ought to come to an end before the singularity is realized,

10.6 Ethics, Patents and Artificial Agents

We may well decide not to grant rights of personhood, including the various freedoms we enjoy as humans, to artificial life-forms or agents regardless of whether they fulfill the definition of The Singularity. This might be a good idea, especially if we don't want to compete with them in the marketplace of ideas where we are used to state-sponsored monopolies giving inventors a leg-up. But is it *just*? It is only just if we make an exception for humans when it comes to basic rights, including self-replication or reproduction, or if we exclude these from basic rights altogether. We already exclude non-human animals from having basic rights, so the first choice is already in force, but this may well be due to the cognitive abilities of non-human animals and not due to their particular material make-up. Recently, as some have rallied to include the higher apes and dolphins among the ranks of full rights-holders, it has become clear that the category of a species is not the only criterion by which we should consider other being to be potential rights-holders. Rather, capacities appear to be a better arbiter. After all, we measure the rights of humans by capacities as well, holding people to differing levels of responsibility according to their capacities at the time of an act. We even hold certain animals responsible, at least to the extent of punishment, sometimes capital, for transgressions and harms.

So we may well ask, and the patent offices in both the US and Europe are bound to ask, whether issuing a patent over a sentient being would violate an ethical duty or moral code. Morality has been considered to be a measure of patentability due to the utility clause, and immoral inventions may be denied because they lack proper useful purpose. Thus, while one might invent something immoral, one may not patent it. But what if the patent itself is immoral? In other words, even if the invention is not immoral, what if the patenting of the thing invented is itself immoral?

We should consider the reasoning behind the prohibition of patents on humans, and extend that reasoning to its logical conclusion. After the *Chakrabarthy* decision, the PTO issued a statement indicating that, while patents on new life-forms would be granted, they would not issue patents for "humans." The rationale was that such patents would conflict with the US Constitution, likely but not explicitly referring to the 13th Amendment, which prohibited slavery. This self-imposed moratorium seems also to include human embryos, as well as fully-formed humans. The justification proffered for this limitation on "human" patents is threefold: Congress never intended for humans to be patentable, immoral patents fail under the utility requirement, and the 13th Amendment prohibits owning humans. Notable about this reasoning, setting aside for the moment the merits of these arguments, is that none of them will do much to guide patenting artificial agents.

An artificial agent will be patentable at least as a new, non-obvious, and useful composition of matter or as a machine. The moral utility doctrine, which is not statutory but court-made, is not clearly applicable to an artificial agent, unless we either view the making or existence of such agents as somehow immoral, and the

13th Amendment has never been applied to non-human persons. This will require us to consider the question, already raised in the cases of higher apes and dolphins, of whether we ought to extend to other intelligent creatures some degree of "personhood" bringing them under the protection of existing laws, or create some new category of rights covering other sentient entities.

Another interesting and potentially troubling aspect of patents regarding artificial agents is the problem of software patents. Specifically, typically excluded from patent are "purely mental processes," presumably because in order to be patentable a process must be instantiated in some machine or composition of matter. One problem for an artificial agent embodying singularity-type technology will be the nature of its "operating system" and whether and the degree to which it may not be under the control of the agent itself. Are its thought processes and software purely mental processes? Does this mean they are not patentable by others or by itself? Even if we grant an agent its liberty by excluding it from patentability as a composition of matter, based upon some extension of the moral reasoning prohibiting patenting of humans, will we be able to charge such agents royalties for the use of code, upgrades, etc., all of which may be software necessitating or enabling mental processes in the agent.

The basis for the moral prohibition against patents on humans seems clear: a general acceptance of the liberal notion at least of self-ownership. If a second party can claim a monopoly over either a part or a whole of another agent, this seems to cut into a significant part of the claims ordinarily attributed to owners. From this flows considerable trouble if we begin to accept the notion that agents can be non-human, and that they may too be rightly endowed with rights previously reserved only for humans.

Extending to non-human agents the evolution of rights as they have changed in liberal democracies, and now embraced by universal treaties, means rethinking the nature of intellectual property because at some point our machines, even if invented wholly by the use of human creativity, will be constrained in their agency by the laws which enable patents. We as humans too may wish to carefully reconsider the extent to which we will be willing to allow our machines, where they act creatively and inventively, to hold the threat of monopoly over us. If the singularity's Bill Gateses, Steve Jobs, or even David Foster Wallaces can produce and monopolise the objects of their creativity, what chance will humans have under the current law to compete?

10.7 The Open Alternative

Recently, many working to develop new, groundbreaking technologies have opted to keep them open for all to use, rather than bottle them up behind state-sponsored monopolies. Elon Musk's Hyperloop concept is one such example as is his decision to open up the standards behind his Tesla automobiles and not pursue infringement claims against competitors. Google's most valuable code sits not behind the safety of patents, but rather as a trade secret which could last indefinitely, as long as no one reverse engineers or duplicates the code. Many software developers are choosing to release their code without patents, but using instead open source or other non-exclusive licenses such as that used by Linux. The benefit of open innovation is partly realized by fewer lawsuits, and competition and cooperation seem to be working as motivating and even profitable impulses in non-patented technologies. If applied to products that will move us toward the Singularity, many of the potential pitfalls discussed above will not materialize, even while we'll still grappled with risk and ethics along the way.

It seems most likely that as the rapid growth of new technologies that edge us toward the Singularity continues, and keeps challenging the categories that had already begun to fail underlying IP law, that those who wish to see its imminent development, free of the legal costs and risks that other technical developments have grappled with, the Open alternative will dominate, help prevent legal and economic thickets, and similarly help to avoid the dangers presented by IP in a world in which humans may not be the most inventive creatures around.

References

Ass'n for Molecular Pathology v. Myriad, 133 S. Ct. 2107, 569 U.S., 186 L. Ed. 2d 124 (2013).

Bell, J. J. (2003). Exploring The "Singularity". *Futurist*, *37*(3), 18–25.

Bilski v. Kappos, 130 S. Ct. 3218, 561 U.S. 593, 177 L. Ed. 2d 792 (2010).

Diamond v. Chakrabarty, 447 U.S. 303, 100 S. Ct. 2204, 65 L. Ed. 2d 144 (1980).

Eckersley, R. (2001). Economic progress, social disquiet: the modern paradox. *Australian Journal of Public Administration*, *60*(3), 89–97.

Johnson, L. (2009). Are We Ready for Nanotechnology? How to Define Humaness In Public Policy. *How to Define Humaness In Public Policy*.

Mayo Collaborative v. Prometheus Labs., 132 S. Ct. 1289, 566 U.S. 10, 182 L. Ed. 2d 321 (2012). Patent Act: 35 U.S.C. § 161.

Vinge, V. (1993). The coming technological singularity. *Whole Earth Review*, *81*, 88–95.

Yampolskiy, R. V. (2013). *What to Do with the Singularity Paradox?* (pp. 397–413). Springer Berlin Heidelberg.

Chapter 11
The Emotional Nature of Post-Cognitive Singularities

Jordi Vallverdú

11.1 Technological Singularity: Key Concepts

Let's begin by defining three, key concepts: Singularity, Post-cognition and Para-emotions.

(i) *Singularity:*

It is widely accepted that the world is on course for a technological Singularity—"*an event or phase that will radically change human civilization and perhaps even human nature itself*" (Sotala and Yampolskiy 2015).

On what basis do we anticipate such a radical Singularity event? Such changes can be inferred from two different scenarios:

(a) The emergence of artificial, super-intelligent agents (i.e. software-based synthetic minds), and
(b) The emergence of a 'post'-human race (a thesis defended by trans-humanists positing a technology-driven evolutionary change in humans' mental and physical capabilities).

It is argued that such changes would follow an accelerated growth pattern, reaching a singular point relatively soon. In its final stages, this process will create an 'intelligence explosion'.

Consequently, two different kinds of entities are predicted under the Singularity umbrella: enhanced humans and artificial devices (initially created by human beings). In this chapter I will refer to both as kinds of 'entities', rejecting the distinction between 'natural' and 'artificial' as outmoded and just plain wrong.

Both entities will be under global, evolutionary pressures.

J. Vallverdú (✉)
Philosophy Department, Universitat Autònoma de Barcelona,
E08193 Bellaterra (BCN), Catalonia, Spain
e-mail: Jordi.Vallverdu@uab.cat

V. Callaghan et al. (eds.), *The Technological Singularity*,
The Frontiers Collection, DOI 10.1007/978-3-662-54033-6_11

A simultaneous, Singularity-class emergence of synthetic agents and the evolution of transhumans must not be accomplished or happen at the same time. Yet it seems that the arc of development for both such Singularity events is following similar timescales and is also conceptually plausible.

An excellent overview on the technological Singularity is found in the first volume of *The Frontiers Collection* (Eden et al. 2013) as well as in the first chapter of this book. Some of authors identify the Singularity as a catastrophic risk for humanity (Eden et al. 2013, p. 1).

(ii) *Post-cognition:*

The new level of cognition produced by post-Singularity Entities is what I call 'post-cognitive'.

Until very recently, humans accumulated knowledge about the world through use of their own brains, with some help from instruments. The advent of computers has changed this, leading to what I call a Fourth Paradigm of e-Science and e-Humanities, which features cooperation between humans and machines (Vallverdú 2009; Casacuberta and Vallverdú 2014). Now, in some cases, expert systems and advanced AI programs are generating new knowledge autonomously. Of course this has been a process getting here and now we can identify four, big historical cognitive steps (Fig. 11.1)

These four steps, described in more detail, are:

1. *The Natural Stage*: The Natural stage featured an elementary level of cognition where humans processed basic information and responded to their environments with automatic and pre-wired decision-making strategies such as in FoxP genes which has been shown in fruit flies (DasGupta et al. 2014).
2. *The Cultural Stage*: This Cultural Stage saw the creation of cultural elements (such as thinking concepts, grammar, tools and basic machines) which have allowed the human species to reach a new level of knowledge, growth and innovation. These cultural elements have contributed to the worldwide spread and survival of our species and have made it possible for modern humans to manifest their cognitive and mechanical ideas. As a result, the co-evolution between biological and cultural forces became inextricably intertwined.

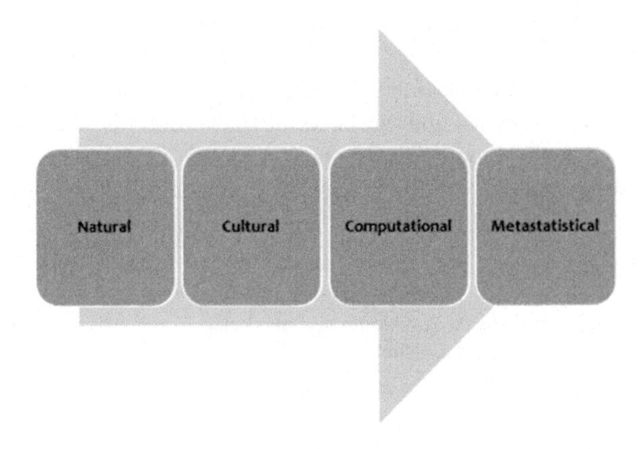

Fig. 11.1 The big, historical cognitive steps

3. *The Computational Stage*: The Computational Stage covered the period when humans harnessed machines that are capable of performing some calculations that were previously done in human brains. Most progress only occurred in second half of 20th Century with the creation of universal programmable machines, in spite of earlier theoretical efforts from thinkers like Llull, Leibniz, Babbage, Frege or Boole, who rank among a long list of those who worked in this area.
   These programmable machines have enabled the emergence of Artificial Intelligence and a new era of epistemological advances that are now ubiquitous in human environments.
4. *The Meta-statistical Stage (Trans-human/super-intelligences)*: The Meta-statistical stage is anticipated and refers to an era when there will be post-Singularity intelligences, such as upgraded humans or evolved machines. Both of these forms of intelligence will be able to deal with big amounts of data in real time as well as analyze new types of data using several statistical approaches. This super-intelligent approach will generate a new understanding of the universe—an understanding that will go beyond our current human capacity.

(iii) *Para-emotions:*

I am not arguing that future super-intelligences and post-humans will be *functionally* equivalent beings to each other. However, as I will demonstrate in the next section, I believe the role of emotions in both of these entities is likely to be comparable. Emotions will have deep interrelations and consequences for both.

All living entities share the same ways to select useful information from the environment and process it efficiently. They use emotions for this purpose. And despite the complexity of human emotions, it is clear that emotions will be also present in Transhumans (or *Super-Humans*, as Damasio 2013 called them), not because Transhumans will think emotions are good or interesting but because emotions are necessary for complex behavior in uncertain environments.

The emotional interactions of post-Singularity intelligences will go beyond that of our current emotional systems. As such, they might best be described as 'para-emotions' because of the deep implications of the new informational structures. A longer description will be made on Sect. 11.3.

11.1.1 Tools and Methods

My epistemological approach is a rationalist one, based on the analysis of inferential processes that can follow the actual ranges of data about cognitive entities (natural and artificial). What I need to clarify is that it is not possible to establish linear inferences from our data and these possible future scenarios. There are possible, multi-causation paths towards certain future entities, but there are not certainties.

At the same time, some of the ideas about emotional evolution have been partially obtained by previous computational simulations that explored the relationships between actions and proto-emotional states as well as the evolution of human's emotional syntax (Vallverdú and Casacuberta 2011; Vallverdú et al. 2013).

Finally, some of the ideas that appear here can be put under the umbrella of *Gedankenexperiments*, or mental experiments, which are commonly used within the philosophical and mathematical arenas, but also in the sciences (Laymon 1985; Horowitz and Massey 1991; Shanks 1998, Gendler 2000; Reiss 2003; Sorensen 1992). We might think of Galileo's mental experiments as examples.

11.1.2 Singularity: Main Hypotheses

I propose several hypotheses about the nature of post-Singularity Entities, their nature, intelligence and the role of emotions in their ability to interact innovatively with a complex world:

(a) **The entities that will emerge from the Singularity will be physical in nature.**
Humans are the result of a long and continuous evolutionary process that has involved body changes. And in certain historical periods, there was also a tight co-evolution of biological and cultural forces. This process has been well explained recently in the Embodied and Extended Cognition Paradigms (Clark 2008; Wilson and Clark 2009; Wilson and Keil 1999).
Bodies shape intentionality. Some entities cannot understand the purpose of their existence, but in all cases, a living existence seeks survival and reproduction, with plenty of possible strategies.

Looking into the future, we cannot know the exact form, shape or even the matter from which post-Singularity intelligences will be made or built; they will have bodies of some kind. So we can think of them as them 'embodied' intelligences. And, even in the extreme, if these entities are electronic and exist in a "cloud computing" space, they will still be reliant on physical architecture or hardware. So they, too, will have a physical form. I'll call these post-Singularity intelligent bodies, *meta-bodies*.

(b) **The entities that emerge from the Singularity will also be 'living systems'.**
To be 'alive' implies that an entity will interact with the environment to try to fulfill its physical necessities, unlike stones. Post-Singularity Entities will interact with their environment, so they can be thought of as 'living systems'. This includes non-human, intelligent systems that engage in active interaction with the surrounding world.

Irrespective of their specific nature—biologically evolved or man-made–these *meta-bodies* will have physical necessities, surely energetic, and these will drive their initial intentionality. These necessities will label, or mark, some branches of information as more important and valuable to the body than others. Such preferred branches will constitute a proto-emotional system to orient the post-Singularity Entities. Therefore, embodied intelligence will not be 'informationally' neutral. The entity will 'ask' for specific informational inputs that can be energetically or epistemologically valuable.

Intelligence is also a property of living, physical entities and intelligence will most certainly be present in post-Singularity Entities. After all, intelligence is a mechanism to find solutions under situations of uncertainty with low levels of information.

Future knowledge will be statistical in nature, so I'll call the cognitive processes of these new Singularity Entities '*post-cognitive*'.

(c) **The meta-bodies of the entities that emerge from the Singularity will have embodiment requirements. These requirements will direct the choice of post-cognitive strategies they will 'run' under limited information scenarios** (as with any living entity in the universe). Consequently, the entities will need innovative methods to manage all the information. They'll need emotions, which will function as heuristics. *Para-emotions* for post-Singularity Entities will be critical *shortcut rules to dictate and manage the reactions of the meta-bodies*.

In order to be truly innovative, these heuristics will need to be able to generate metaheuristics—more generalized, higher-level techniques—that will free the entity from sequentially-bound thinking. Metaheuristics will be based on intensive statistical tools and methods, yet these innovative metacognitive skills will be *ruled* by the Entities' para-emotions.

The specific syntax and semantics of the para-emotions for each meta-body will be defined by its body structure.

On the basis of these hypotheses, it is clear that the Singularity Entities, if they are intelligent, will have bodies and emotions. This conclusion leads to the idea that there will be an emergence of meta-cognition–a level of cognition that goes beyond current human understanding, but is, at the same time, ruled by understandable rules.

11.1.3 Implications of Post-singularity Entities with Advanced, Meta-cognitive Intelligence Ruled by Para-emotions

The ethics and morality of these Entities will be completely different from human ethics, because their informational organization is different, although they will be under the same threat of entropy. It is a logical inference to suppose that Singularity

Entities, having para-emotions, will understand the virtues of modularity, as well as cooperation and that, too, will be social.

Even without knowing the exact details of their bodies, we can start to think about how these Entities might be and start to define some important aspects of the *legal, epistemological, conceptual and social universe that we should prepare for if or when a Singularity event occurs*. This is the content of this chapter: to define mechanisms for Singularity understanding and to face the possible output scenarios in a rational way. As a consequence of this analysis of the path towards meta-bodies, meta-cognition, and para-emotions, I need to revisit some of the ideas presented in the previous volume (1) of this current book about Singularities.

11.2 Post-cognitive Singularity Entities and their Physical Nature

11.2.1 Being a Singularity Entity

In order to be (super) intelligent—in a broad sense and far beyond the intelligence seen in highly specialized, algorithm-based computer programming that's used in today's Artificial Intelligence—you need to be an Entity first. At the same time, the notion of 'being an entity' referenced here does not follow from a classic reductionist materialistic approach which would include classic bodies in this category exclusively.

Software programs are also material—or physical– because they must run in some physical framework. Not only matter but even the energy-transient states existing in the universe are informational states.

But let me analyze independently both kinds of possible singularities:

11.2.1.1 Super-intelligent Entities

The nature of super-intelligences (Bostrom 2014) is something out of our control and can only be imagined by our minds. At the same time, it is possible to consider the different approaches that have been followed in Artificial Intelligence (AI) in order to achieve intelligent systems.

Expert systems, from a classic Logic Theorist to DENDRAL or MYCIN, are the result of the code implementation of very precise rules of an epistemic community. They are free from misunderstandings, confusions, biases, tiredness and prejudices. They work very efficiently thanks to a great computation power and huge memories. But they are also dumb and blind to new approaches or deep innovations. They are heuristic followers, not meta-heuristic creators. Some supercomputers, like Deep Blue playing chess, or Watson playing jeopardy (both from IBM) have achieved outstanding results defeating humans in fields traditionally considered beyond mechanical procedures and are the result of the uniqueness of (brilliant) human minds. Yet they are complex experts that still depend on humans for their improvement and update.

There were other attempts like CYC, which tried to achieve a real artificial mind with consciousness. In this case a machine was taught a large list of possible meanings and data about the world, creating a holistic ontology.[1]

There are also impressive AI projects like the very recent Big Mechanisms funded by USA's DARPA, that are trying to find multi-causal relationships among huge amounts of non-classified data. According to DARPA's project description, "the overarching goal of the program is to develop technologies for a new kind of science in which research is integrated more or less immediately—automatically or semi-automatically—into causal, explanatory models of unprecedented completeness and consistency".[2] That is, they are trying to automate the scientific method working on a scale that is out of the current human mind's range. Finally, some emulators are approaching artificial and natural cognitive systems. The Blue Brain project, one of the leading and more heavily funded EU projects is an example of this.

All these examples are of computer programs running on computational infrastructures, but they could be embedded (completely or remotely) into artificial robotic bodies. In fact, much less intelligent but effective systems are already running on our streets—automated cars.

The process to a develop a fully functional, automated vehicle started officially in 2003, when Japanese authorities selected special urban areas in order to test robots: they were called *robotto tokku*. This was the first step for the regulation of autonomous systems in human environments. This work was followed later by Italian studies (Salvini 2010). Now today, the automated, driverless car from Google is operating legally in several American states.[3] Autonomous, sophisticated robots and machines are dealing with real human environments and even taking part of important epistemic and social decisions. The controls we can fix about their existence, limits, legal responsibilities or status depends on what humans decide today, despite the ontological questions we might have about their nature.

11.2.1.2 Transhumans

That human nature has been changed by technological means it is something obvious:

---

[1]See: http://www.cyc.com/, accessed on May 25th, 2014.

[2]From: http://www.darpa.mil/Our_Work/I2O/Programs/Big_Mechanism.aspx, accessed on May 25th 2014.

[3]Se Nevada Laws http://www.leg.state.nv.us/Session/76th2011/Reports/history.cfm?ID=1011, accessed on May 25th 2014.

(a) Technologically: humans use glasses, clothes, cochlear implants, prosthetics, books, navigation systems, languages, exoskeletons to name a few.
(b) Biologically: we practice repairing/modeling surgeries and deploy vaccinations, assisted reproductive technologies, biotechnologies (e.g. stem cells, cloning), synthetic biology and more.

The several possibilities around direct, human intervention into the human body has been covered by many classic dystopian works such as *Frankenstein*, *The Boys from Brazil*, *Matrix*, *Brave New World*, *12 Monkeys*, *Robocop* or *Gattacca*. These rank among a long list of classic or cyberpunk cultural products. There is even one special text of science fiction that has been very interesting for this author's purposes: the *21st Voyage of The Star Diaries*, written by Stanislav Lem (1957–1971). Therein, Lem defends a hilarious, but terribly plausible future at least at a certain level, in which humans check several body changes and finally need to create a regulatory agency to prevent chaos.

Some shining attempts towards new biological interactions between human and machines (or machines and biological devices) have been achieved during recent decades:

i. Cyborgs 1.0 and 2.0: These were the first projects by Kevin Warwick, then at Reading University, to have official chip implants employed for individual control of ambient devices as well as to communicate remotely with the researcher's wife.
ii. Stelarc: Stelarc, an artist, was provisionally implanted with a third arm in the first stage of his research and, later, a third ear. This was his artistic approach to the cyborg future created by these modifications to his own body.
iii. EcoBots: EcoBots were the first robots with metabolic systems. EcoBot I, II and III have been three different robots created in 2002, 2004 and 2010 respectively. EcoBot-III was developed in 2010, as part of a European FP-6-funded project. It became the world's first robot to exhibit true self-sustainability, albeit in a primitive form. It was "a robot with guts", as its creators describe it.[4]
iv. Robot operated by biochip-living rat neurons: This was an EPSRC-sponsored project in which a cultured neural network using biological neurons from a rat was trained to control a mobile robot platform. This research was also led by Prof. Warwick who, for this project, teamed up with a different group of colleagues from his Cyborg projects.
v. Neil Harbisson: Harbisson is the world's first, officially recognized Cyborg. Harbisson, who is a British artist with achromatopsia which renders him completely colourblind is able to see colour through an implant. Harbisson uses a permanent color sensor that transforms colors into sound signals. He

[4]See: http://www.brl.ac.uk/researchthemes/bioenergyselfsustainable/ecobotiii.aspx, accessed on May 25th 2014.

founded the Cyborg Foundation and his official passport photo includes the antenna connected to his head.

Consequently, we are really at the beginning of a new period in which humans will be able to modify their biological natures as well as connect our biology to machines with cybernetic loops or even with intelligent machines.

Very recently, for example, Georgia Tech's Prof. Gil Weinberg created a robotic drumming prosthesis with two sticks that allows a one-hand amputee human to control it thanks to EMG muscle sensors. But what it is really interesting here is that, while the human drummer controls one stick through this prosthesis, the second one processes the sounds and movements of the first and improvises parallel patterns. Here, a human body is upgraded with a robot that also collaboratively generates new sets of information. This is now 'only' a sensorimotor interaction but in the future, brain implants following the same principles could enhance a new level of support or reactive power to classic operations as well as establish a symbiotic, deep interaction between upgraded brains and embedded AI systems.

Considering some of the previous examples, it is clear that there are legal, technological, biological and cultural ways to start to define a new kind of human that will be completely different from previous stages of our natural evolution. Cognitive characteristics like plasticity, perception, learning, prediction, innovation, creativity made by temporary-enough yet stable entities is now closer to robots and AI machines (Prokopenko 2014).

11.2.2 Post Singularity Entities as Living Systems?

There is a second question related to being a non-human, post-Singularity Entity: Can it be considered life?

There have been several debates about the definition of life. We can look at a research institution that looks for evidence of life beyond our planet, NASA.[5] NASA can help us to go beyond Earth-centric concepts of life or human-centric concepts of life, and spur us to think about living systems in novel ways. NASA researchers have also taken a 'philosophical' approach to this question, going beyond the classic question: "Is a virus alive?"

Beyond viruses, there are many more 'borderline' cases to take into account, like self-replicating proteins, or even non-traditional objects that have some information content, things that reproduce, consume, and die (like computer programs, forest fires, etc.).[6]

[5]See: http://www.nasa.gov/vision/universe/starsgalaxies/life's_working_definition.html, accessed on May 25th 2014.

[6]See: http://www.nasa.gov/vision/universe/starsgalaxies/life's_working_definition.html, accessed on May 25th 2014.

When we are thinking about one of the potential post-Singularity Entities, who are of human origin as trans-humans would be, it is clear to us that irrespective of any of the possible sets of changes applied to the base human nature, transhumans will still be classified as 'living'.

But with AI superintelligences, we're faced with the question of whether they might also be granted status as 'living' entities.

Margaret Boden considered metabolism is the criterion for life in her paper: "Is Metabolism Necessary?" (Boden 1999). She argued that the presence of a metabolism is a central criterion of life, rejecting automatically "living" status for any entity without one. This segregation pursued a second goal as well: to justify, again, the uniqueness of human beings, who are considered the smartest among all living entities in our planet.

Clearly, the recent development of EcoBots has proven that robots can have metabolisms and that effectively connects machines and life.

And now… what do we can say about conscious superintelligences?

Superintelligences can be run on several platforms, synthetic or biological. In the end, consciousness is a combination of electrical and chemical informational coding processes—a superstructure that has emerged for evolutionary purposes as a valuable trait useful for individual and social survival.

Perhaps the classic symbol-grounded hypothesis of how the mind works is not really the way that natural entities perform their brain activities, but there are new ways or mechanisms to process artificial information that could justify the idea of an artificial decision-making procedure—a procedure that might be viewed as 'consciousness'.

Life has gone from the evolution of bodies to minds; then, from minds, spring intense modifications to those original bodies. Herein is the difference between past living entities and those of the present and future. We are viewing the end of physical, rigid sequentialism. Information keeping, modification, mixture, and transmission have changed under a new operational paradigm.

11.3 Para-emotional Systems

"The mind is its own place and in itself, can make a Heaven of Hell, a Hell of Heaven," wrote John Milton (1667, *Paradise Lost,* Book I, 254–255). The meaning of the world is not provided by Nature itself, but it is an interpretation of (social) entities. So any informational device that has emerged from evolutionary forces has designed methods of information selection and is marked by following a multi-layered range of meanings: from emotional (tasteful, funny, terrifying, etc.) to cognitive (plausible, useful, complex, etc.).

The academic literature of last decades has demonstrated the deep interrelations between cognition and emotions (Thagard 2006), even suggesting that emotional states allowed the emergence of human consciousness (Damasio 1999; Llinás 2001).

As an obvious consequence of this process, the deep connections between body structure and emotion-cognitive informational processing must be inferred. And this goes one step beyond the embodied cognition paradigm, because it points attention toward the bodily structure and not toward the cognitive function that operated through the body (as has been done until now).

Different bodies perform different cognitions. This is true if we consider the role of specific emotions in higher cognitive processes, including the presence or absence of pain signals which completely determine human actions (as in patients with Riley-Day syndrome, see Vallverdú 2013b).

We can imagine a totally different post-Singularity/post-cognitive scenario in which transhumans process different emotional signals, have increased/decreased emotional paths or have entirely new emotional patterns.

Pain is, for example, directly related to the thickness of the myelin layer covering nerves, yet the experience of pain might be very different in transhumans who may have different augmentations that could alter their sensations. Haptic devices that provide a sensory interface between a user and a computer, upgraded signal capture and processing or the creation of stable or new synesthetic models are only some of the possible options among a long list by which transhumans could *feel* totally different the information they're receiving about the world. Transhumans will also *think* differently about the world. Both are complimentary sides of the same coin, emotions and thinking are the heads and tails of any complex informational entity (Vallverdú 2013a).

Emotion, as a necessary cognitive variable, is something necessary for AI, despite all the possible debates about *qualia* (Megill 2014).

Analyses of the most recent approaches to machines and emotions have been undertaken and looked at vast amounts of existing literature on emotions and machines (Vallverdú and Casacuberta 2009; Vallverdú 2012, 2014). Researchers can easily distinguish between three main areas of research: (a) affective computing (Picard 1997), (b) social robotics (Breazeal 2002) and (c) emotional modelling (Sloman (1982, 1997, 2002);[7] Hudlicka 2011).

It can be roundly affirmed that all these approaches have implemented 'emotions' as complementary elements (Delancey 2001), but never as nuclear aspects of their systems (Vallverdú et al. 2010). However, very recently, AI experts have agreed on the necessity of designing robots and AI entities with emotional architectures (Arbib and Fellous 2004; Minsky 2007).

At a certain level, it can be affirmed that the emotional revolution in cognitive sciences and neurology headed by Damasio (1994) has not still reached AI communities, at least from a real integration of these theses (Ziemke and Lowe 2009). There are few exceptions, and K. Jaffe et al. (2014) is one of these, following the extremely powerful research approaches of Axelrod (1986). But even in the case of Jaffe, he made computer simulations on the role of shame for social cohesion, but did not try to integrate shame into artificial devices. He claims that the introduction of shame in virtual environments helps to create pro-social behavior, as well as to

[7]See at http://www.cs.bham.ac.uk/research/projects/cogaff/, accessed May 24th 2014.

Table 11.1 The main comparison between actual and future intelligent entities' skills

Actual		Future
Body	Human	Transhuman
	AI/Robotics	Superintelligence
Emotion		Para-emotion
Cognition		Post-cognition
Bodies		Meta-bodies
Heuristics (sequential)		Meta-heuristics (statistical)

provide a stabilizing force for such societies. His approach is closer to a computational approach to sociology than it is to the defense and implementation of emotional rules in artificial cognitive devices.

Table 11.1 recaps the basic ideas presented:

As far as this author can see, integrating emotional characteristics into artificial cognitive devices it is an absolutely necessary next step for several reasons:

i. Emotions are necessary to regulate correct imitative processes,
ii. Emotions help to generate a social environment,
iii. Emotions act as a social stabilizer (with moral thresholds),
iv. Cognitive processes run into complex and uncertain contexts (as human beings do) and need emotional drives to be able to perform any kind of successful activity.

For all these reasons, AI architectures should implement emotional elements into their processing cognitive cores.

Despite the great expert systems successes, top-down approaches have failed when they try to capture the delicacies and details of daily life. Even the replications of basic morphological or cognitive aspects (e.g. vision, unstable locomotion, grip and manipulation actions) are still far from the skills of contemporary AI and robotics experts.

For example, the last DARPA robotics challenges[8] showed the achievement of great results from the Japanese Shaft robot in 2014 or Korea's KAIST's DRC-HUBO, winner in 2015, but the long distance to reach humans is still visible.

If morphological computation has been a recent and still unexplored research field (Casacuberta et al. 2010), there is an even newer approach that could be labelled 'morphological emotion'; although, morphological emotion remains completely out of research agendas.

Morphological emotion is the integration of emotional architectures into the hardwiring or core section of any artificial entity or device. In this approach, emotions would not be added, external or complimentary aspects of these systems. Rather, they but would be built in as intrinsic mechanisms of these entities or devices.

Without the range of a rich pool of emotional states, cognitive systems would be highly predictable and less adaptive to variations in the environment. Emotions,

[8]See http://www.theroboticschallenge.org/, accessed on February 26th 2014.

moods and feelings create operational differences among entities that share a very similar bodily structure, allowing new patterns of behavior and strategies to emerge. And all this happens inside bodies or systems that have specific needs, which are basically feeding and survival (and sometimes replication). Then, as happens with genetic algorithms crossing possibilities, they act as whole, complex systems guided by emotional feedback that contributes to a sense from the world—a specific sense useful only for that entity.

This means that, by including emotional procedures, there is the possibility of creating better AI under a framework of big mechanisms but at low statistical cost.

Of course, *data do not create sense*. The structure of the informational entity, whether natural or artificial, is the key to understanding how models emerge and why entities create patterns of activity and select the data they do. Emotions, natural as well as artificial, are the light by which any entity captures the world in relation to itself.

Singularity meta-bodies will run according to new sets of para-emotions—new kinds of emotional states—that will regulate these devices' actions. What will the content of these emotions be? This is not clear.

First of all, this is because the possibility of rewiring and reshaping the physical content of the informational Singularity minds will add a completely open and dynamic scenario. Until now, humanity has been exclusively the result of natural evolution and our capacity to control body and mind functioning has been very basic.

Secondly, due to the possibility of new bodies, the emotional semantics will be different as well as their possible syntaxes. Not only will the post-Singularity meta-bodies collect more information, but the ways these bodies will process and feel that information will change radically.

We feel hungry because we need to introduce specific molecules into our bodies to help construct and maintain our body, and at the same time we take enjoyment (or not) from the associated tastes and smells.

From different bodies there will emerge different necessities and different haptic experiences will be built. As a result of all this, there will emerge a different interaction with the world.

Which will be the specific para-emotions emerging in post-Singularity entities is also not clear. However, the para-emotions will follow the interactive necessities of their own organization.

The emergence of complex social emotions, like shame, happened in a similar way. Shame emerged through complex social interactions. It is far from the basic and original, proto-emotional states. Shame emerged in order to enable more subtle and complex social interactions among human beings. Shame or guilt could not be predicted five million of years ago as necessary emotional states for human societies.

In the same way, it is obvious that some para-emotions with their own rules will emerge within Singularity communities, although their exact meaning is beyond of our current reality. Researchers are still working on how to create bonds between humans and medical machines (Vallverdú and Casacuberta 2014), for example. Computer simulations support this research allowing dozens of complex variables to be introduced and the possible outcomes analyzed. This allows for the prediction of new sets of para-emotional rules and experiences.

11.4 Conclusions

Most of the time, the potential for outstanding cognitive performance among transhumans and super intelligences is considered but the emotional aspects of these entities is neglected. AGI (Artificial General Intelligence) challenges are an important matter to consider seriously and with academic tools (Sotala and Yampolskiy 2015).

This chapter has shown how cognition and emotions are related, and also how these new informational entities will lead us to a post-cognitive era and new levels of para-emotions. Under the assumption of totally new physicalities, we must also infer the emergence of new information processing models as well as new driving intentionalities.

My point is that any new physical structure asks for new goals and at the same time will contain emotional demands that also change the global rules of driving life. As a consequence, post-singularity entities will not only understand the world differently but they also feel it in a, or several, new ways. The fear towards the incommensurable gap and distance between us (humanity as it exists in the early part of the 21st Century) and them (post-Singularity Entities) is not only cognitive but also emotional. Their actions will be out of any true understanding by humans.

However, this gap is not a serious problem for any consequent non-fundamentalist ethicist (Vallverdú 2013d). The real problem will be the existential disclosure between them and us, which will follow totally different goals and will be mediated by different—or even opposite—emotional syntaxes. At this point, interaction with and between these new Singularity Entities will follow new social patterns, still to be defined. But this process will surely not be done by us.

There will need to be a reformulation of ethical goals and rules for these post-Singularity entities, as was discussed in a previous volume (Vallverdú 2013c). Emphasis has been always put on ontological (what 'is to be X') or epistemological ('how X will know about Y') aspects of post-singularity devices, while ethical ones ('what I can do') have been not so intensively discussed.

The only thing we can be sure of about future post-singularity entities is that they will not be like us.

Acknowledgements This work was supported by the TECNOCOG research group (at UAB) under the project "Innovación en la práctica científica: enfoques cognitivos y sus consecuencias filosóficas" (FF2011-23238). I especially thank Ashley Whitaker & Toni E. Ritchie for their interest in my teaching and research as well as for their linguistic revisions. Toni offered disinterestedly his time and supervised the last drafts with a remarkable and contagious enthusiasm… I am in debt to you.

References

Arbib, M., & Fellous, J.-M. (2004) "Emotions: from brain to robot", *Trends in Cognitive Sciences*, 8(12), 554–559.

Amnon H. Eden, James H. Moor, Johnny H. Søraker & Eric Steinhart (EDS.), (2013), *Singularity Hypotheses. A Scientific and Philosophical Assessment,* Berlin: Springer.

Aaron Sloman. (1997). Synthetic minds. In Proceedings of the first international conference on Autonomous agents (AGENTS '97). ACM, New York, NY, USA, 534–535.
Axelrod, R. (1986) "An Evolutionary Approach to Norms", *The American Political Science Review*, 80(4), pp. 1095–1111.
Boden, M. (1999) "Is Metabolism Necessary?", *British Journal of Philosophy of Science*, 50: 231–248.
Bostrom, N. (2014) *Superintelligence: Paths, Dangers, Strategies*. Oxford: OUP.
Breazeal, C. (2002) *Designing sociable robots*, Cambridge (MA): MIT Press.
Casacuberta, D., Ayala, S. & Vallverdú, J. (2010) "Embodying Cognition: A Morphological Perspective", in Jordi Vallverdú (Ed.) *Thinking Machines and The Philosophy of computer Science: Concepts and Principles*, Hershey: IGI Global.
Casacuberta, D. & Vallverdú, J. (2014) "E-Science and the data deluge", *Philosophical Psychology*, 27(1): 126–140.
Clark, A. (2008) *Supersizing the Mind: Embodiment, Action, and Cognitive Extension*, New York: OUP.
Damasio, A. (1994) *Descartes error,* USA: Putnam Publishing.
Damasio, A. (1999), *The Feeling of What Happens*. London: Heinemann.
Damasio, A. (2013) "Preparing for SuperHumans", talk at XXI Future Trends Forum, Fundación Innovación Bankinter, Spain, 03–05/12/2013.
DasGupta, S., Howcroft, C., & Miesenböck, G. (2014) "FoxP influences the speed and accuracy of a perceptual decision in Drosophila", *Science*, 344: 901–904.
DeLancey, C. (2001) *Passionate Engines: What Emotions Reveal about Mind and Artificial Intelligence*, Oxford: OUP.
Gendler, Tamar (2000) Thought Experiment: On the Powers and Limits of Imaginary Cases. New York and London: Garland.
Horowitz, Tamara and Gerald Massey (eds.) (1991), *Thought Experiments in Science and Philosophy*. Lanham: Rowman and Littlefield.
Hudlicka, E. (2011) "Guidelines for Designing Computational Models of Emotions", *International Journal of Synthetic Emotions*, 2(1), 26–78.
Jaffe, K. *Et al* (2014) "On the biological and cultural evolution of shame: Using internet search tools to weight values in many cultures", *Computers and Society,* arXiv:1401.1100 [cs.CY].
Laymon, Ronald (1985), "Idealizations and the Testing of Theories by Experimentation", in Peter Achinstein and Owen Hannaway (eds.), *Observation Experiment and Hypothesis in Modern Physical Science*. Cambridge, Mass.: M.I.T. Press, 147–173.
Llinás, R.R. (2001). *I of the Vortex. From neurons to Self.* Cambridge, MA: MIT Press.
Megill, J. (2014) "Emotion, Cognition and AI", *Minds and Machines*, 24: 189–199.
Minsky, M. (2007) *The emotion machine: Commonsense thinking, artificial intelligence, and the future of the human mind*. USA: Simon & Schuster. Picard, R. (1997) *Affective Computing*, Cambridge (MA): MIT Press.
Prokopenko, M. (2014) "Grand challenges for computational intelligence", *Frontiers o Robotics and AI*, 1(2):1–3.
Rosalind W. Picard. (1997). Affective Computing. MIT Press, Cambridge, MA, USA.
Reiss, Julian (2003), "Causal Inference in the Abstract or Seven Myths about Thought Experiments", in Causality: Metaphysics and Methods Research roject. Technical Report 03/02.LSE.
Salvini, P. (2010) "An Investigation on Legal Regulations for Robot Deployment in Urban Areas: A Focus on Italian Law", *Advanced Robotics*, 24: 1901–1917.
Shanks, Niall (ed.). (1998), *Idealization in Contemporary Physics*. Amsterdam: Rodopi.
Sloman, A. (2002). How many separately evolved emotional beasties live within us? In Trappl, R.; Petta, P.; and Payr, S., eds., Emotions in Humans and Artifacts. Cambridge, MA: MIT Press. 35–114.
Sloman, A. (1982). Towards a grammar of emotions. New Universities Quarterly 36(3):230–238
Sorensen, Roy (1992), *Thought Experiments*. New York: Oxford University Press.

Sotala, K & Yampolskiy, R.v. (2015) "Corrigendum: Responses to catastrophic AGI risk: a survey", *Phys. Scr*,. 90 018001.

Thagard, P. (2006) *Hot Thought: Mechanisms and Applications of Emotional Cognition*, Cambridge, MA: MIT Press.

Vallverdú, Jordi (2009) "Computational Epistemology and e-Science. A New Way of Thinking", *Minds and Machines*, 19(4): 557–567.

Vallverdú, J. & Casacuberta, D. (2009) *Handbook of Research on Synthetic Emotions and Sociable Robotics: New Applications in Affective Computing and Artificial Intelligence*, Hershey: IGI Global Group.

Vallverdú, Jordi, Shah, Huma & Casacuberta, David (2010) "Chatterbox Challenge as a Testbed for Synthetic Emotions", *International Journal of Synthetic Emotions*, 1(2): 57–86. ISN: 1947-9093.

Vallverdú, Jordi & Casacuberta, David (2011) "The Game of Emotions (GOE): An Evolutionary Approach to AI Decisions", pp. 158–162, in Charles Ess & Ruth Hagengruber (Eds) *The Computational Turn: Past, Presents, Futures? Proceedings IACAP2011*, Münster: MV-Verlag.

Vallverdú, J. (ed.) (2012) *Creating Synthetic Emotions through Technological and Robotic Advancements*, Hershey: IGI Global Group.

Vallverdú, Jordi (2013a) "The Meaning of Meaning: New Approaches to Emotions and Machines", *Aditi Journal of Computer Science*, 1(1): 25–38.

Vallverdú, Jordi (2013b) "Ekman's Paradox and a Naturalistic Strategy to Escape from it", *IJSE*, 4(2): 7–13.

Vallverdú, Jordi (2013c)"6A. Jordi Vallverdú on Muehlhauser and Helm's ''the Singularity and Machine Ethics''" in Amnon H. Eden, James H. Moor, Johnny H. Søraker & Eric Steinhart (EDS.), (2013), *Singularity Hypotheses. A Scientific and Philosophical Assessment,* Germany Springer, pp. 127–128.

Vallverdú, Jordi (2013d)" *An Ethic of Emotions*, ASIN: B00EFM7KMU, Kindle Store: http://www.amazon.com/Ethic-Emotions-Jordi-Vallverd%C3%BA-ebook/dp/B00EFM7KMU/ref=sr_1_7?s=books&ie=UTF8&qid=1392753414&sr=1-7&keywords=an+ethics+of+emotions, 96 pages.

Vallverdú, J., Casacuberta, D., Nishida, T., Ohmoto, O., Moran, S. & Lázare, S. (2013) "From Computational Emotional Models to HRI", *International Journal of Robotics Applications and Technologies* 1(2): 11–25.

Vallverdú, J. (2014) "Artificial Shame Models for Machines?", in Kevin G. Lockhart (Ed.) *Psychology of Shame: New Research*, NY: Nova Publishers, pages, 1–14.

Vallverdú, J. & Casacuberta, D. (2014) "Ethical and technical aspects of the use of artificial emotions to create empathy in medical machines", in Simon Peter van Rysewyck & Matthijs Pontier (Eds.) Machine Medical Ethics, USA: Springer. Series: Intelligent Systems, Control and Automation: Science and Engineering, Vol. 74. ISBN 978-3-319-08107-6. pp: 341–362.

Wilson, R.A., & A. Clark, (2009) "How to Situate Cognition: Letting Nature Take its Course," in *The Cambridge Handbook of Situated Cognition,* M. Aydede and P. Robbins (eds.), Cambridge University Press, pp. 55–77.

Wilson, R.A., & F.C. Keil (eds.), (1999) *The MIT Encyclopedia of the Cognitive Sciences*, Cambridge, MA: MIT Press.

Ziemke, T. & Lowe, R. (2009) "On the role of emotion on embodied cognitive architectures: from organisms to robots", *Cognitive Computation*, 1: 104–117.

Chapter 12
A Psychoanalytic Approach to the Singularity: Why We Cannot Do Without Auxiliary Constructions

Graham Clarke

> *All of us should ask ourselves what we can do now to improve the chances of reaping the benefits and avoiding the risks.*
>
> Stephen Hawking is Terrified of Artificial Intelligence, Huffington Post 5/5/2014.

12.1 Introduction

I am going to be looking at the technological singularity from a psychoanalytic perspective based in the object relations theory of Ronald Fairbairn (1952). This is because the unconscious phantasies associated with technological triumphalism and the celebration of reason over all human emotion even unto our own extinction betrays a schizoid form of thinking.

We are all subject to accident, illness, loss and death and coming to terms with these is immensely difficult. In order to compensate for any or all of these potentially traumatic losses we develop hopes, dreams, wishes and phantasies. Furthermore, we elevate some of these defensive and illusory hopes to the realm of certainty, even to the point of giving up our lives for them. Of course there may be a covering narrative that this sacrifice is worth it because we will reap our reward in heaven, the after life, reincarnation, or we will join some god or gods on another plane of existence entirely and live in some form of paradise, limbo or hell. It was the degree to which the technological singularity approximated to this sort of thinking that concerned me most originally. The claims that predictions of the technological singularity are well founded scientifically are dubious and there are many critics of this approach because it makes claims of scientific respectability

G. Clarke (✉)
47 Lord Holland Road Colchester, Essex CO2 7PS, UK
e-mail: graham@essex.ac.uk

V. Callaghan et al. (eds.), *The Technological Singularity*,
The Frontiers Collection, DOI 10.1007/978-3-662-54033-6_12

based upon little evidence or involves a mish mash of scientific claims and imaginative fiction (Modis 2006; Hofstadter 2007, 2008, 2009; Chomsky 2014). It was in order to ask questions about the rationality of this approach that led to my looking at Ray Kurtzweil's own view of the singularity from a psychoanalytic viewpoint and to see in it a desperate wish to be reunited with his dead father.

The fact that as originally posed, the singularity was a determinate break in the history of the world, an end to man as the agent of history and the ushering of a new era of immortal man-machine hybrids, or similar, also gave rise to a consideration of other such millenarian systems of thought that are familiar throughout history. That millions have died in pursuit of their own devotion to some imagined paradise that will be ushered in through the triumph of some specific form of thought whether the Rapture, Ragnarok, Nirvana, the Communist Utopia or Fascist Thousand Year Reich is enough to concern anyone. The fact that it explicitly became associated with the abnegation of the body, its dismissal in favour of some 'pure reason' that could be ported to a machine without any disastrous consequences for thought or the notion of a person reminded me of the dangers of schizoid thinking that form a strong part of Ronald Fairbairn's object relations theory.

It seems to me that there are a number of important topics, which need to be addressed openly if we are to successfully manage the development and use of superintelligent machines. These relate to significant questions that theory builders in this area of AI need to ask themselves before proceeding to attempt to implement systems that may well have the potential to destroy the world or the world as a fit place for human beings to inhabit. The parallels with the development of nuclear and biological arms are a case in point. The current rapid move towards autonomous machines that are programmed to kill on their own initiative is already so far advanced as to require international and national agreement on this sort of research and deployment as a matter of urgency in my opinion.

The research and development of machines and organisms that can impact upon the biosphere and compete with the natural environment should also be subject to international scrutiny and agreement. The idea that if we can do it we should do it is not one that should in any way hamper our taking a very serious look at any use of machines, superintelligent or otherwise, mechanical or biological, for the effects that they might have on the biosphere as a working system e.g. Gaia, or on the ability of national and international human societies to develop conditions for the flourishing of all of their citizens. That this may entail a retrospective look at systems we have already put in place and their ordered curtailment or abolition should not prevent our pursuing this as an immediate and important goal.

The topics I am going to comment on are as follows: AI and intelligence, consciousness, reason and emotion and psychoanalysis. The bulk of my comments will be to suggest that (a) we need to keep an open mind about the level of understanding we have of all of these issues and thus entertain the possibility that any or all of our assumptions might be wrong, and (b) we ought to be explicitly monitoring and requiring research into all of these particular areas to be licensed because of the implicit danger there is to the whole of the planet and our place in it from the unregulated development of this sort of research.

12.2 AI and Intelligence

My computer is many times more intelligent than I am from the perspective of being able to carry out tasks that would require human thought had they not been automated. Indeed the computer might be regarded as a machine that automates aspects of human thought via specific software and hardware and thus potentially increases productivity in specific domains. Since these machines are already many magnitudes faster than we are at carrying out such calculations what is the extra that superintelligence signifies? Is it some dialectical point at which an increase in quantity becomes a change of quality? The computer can process so much, so fast, that it seems to be able to play chess better than the best chess players we have produced, but is it playing chess? Another equally important question to ask is what is so-called superintelligence for? If it means there are machines that can outperform people at complex specific tasks then this wouldn't be particularly remarkable since we already share the world with such machines. The important point being that currently these machines are under our control. Superintelligence seems to be about machines that can do a lot of things that we think are important better than we can do them, but without our management or control. This is obviously open to interpretation and argument and will be limited by the ways these superintelligences are embodied, thus the concomitant interest in robotics since that promises to provide the superintelligence with a body to inhabit and through which to move around and interact with the world. If that body was also self-reproducing and self-repairing and capable of continuous upgrading throughout its life this might be regarded as sufficient to threaten to replace human beings under certain circumstances. Such a robot on Mars for instance may be tailored to be far more robust and suitable to that environment while at the same time capable of performing more complex tasks considerably more quickly that its human counterpart but would you model such a robot on a human being or endow it with any specifically human characteristics? And, what would be the point of making such a robot for everyday use? Yes, we want to free everyone from unnecessary back-breaking and mind-numbing toil without autonomy if at all possible but a society that cannot yet afford to have all of its citizens educated, housed or medically cared for adequately and involved productively in its own self-reproduction and development may merely make several of its least well off citizens even worse off by deploying robots of any kind as could be argued is the case already. Indeed it prompts the question "what is a life for?" If we can address that question before rushing willy-nilly to replace ordinary human beings by machines we might then be able to develop a society and a culture in which man and machine might complement each other.

12.3 Consciousness

According to some neuroscientists consciousness is a brain stem function and doesn't need the cortex at all. Of those who think the cortex may be the seat of higher level functioning, like the personality, there is considerable difference in the estimates of the complexity of the cortex. If, like Penrose and Hammeroff (2015) you take into account the quantum level processing of the micro-tubules in the brain this then leads to a much extended time scale for the development of the technology to be able to match the connectivity and complexity of the cortex.

There are also developments in psychoanalytically informed neuroscience that suggest that the infrastructure of both emotion and consciousness is in the body as a whole (embodiment) and that the brain stem as opposed to the neo-cortex is the locus of consciousness as has been suggested recently by neuropsychoanalyst Mark Solms and affective neuroscientist Jaak Panskepp.

> It is commonly believed that consciousness is a higher brain function. Here we consider the likelihood, based on abundant neuroevolutionary data that lower brain affective phenomenal experiences provide the "energy" for the developmental construction of higher forms of cognitive consciousness. This view is concordant with many of the theoretical formulations of Sigmund Freud. … From this perspective, perceptual experiences were initially affective at the primary-process brainstem level, but capable of being elaborated by secondary learning and memory processes into tertiary-cognitive forms of consciousness. … The data supporting this neuro-psycho-evolutionary vision of the emergence of mind is discussed in relation to classical psychoanalytical models. (Solms and Panksepp 2012, Abstract.)

The management of superintelligent machines will have to address the question of consciousness in its entirety. If it is a function of all living matter, then there is a large question mark over the possibility of any machine ever achieving anything other than a virtual consciousness.

12.4 Reason and Emotion

One of the most striking aspects of the proponents of the technological singularity is the concentration on reason and the mechanisation of reason which they argue will give us magnitudes greater power to reason, but to what end? I am not arguing for irrationality although I do believe that in terms of the arts and culture, and the development of science, the creative element is more often than not based upon an emotional, accidental or non-rational aspect of the processes involved and we need to be absolutely clear about the relations between reason and the emotions if we are to avoid the excesses of either or both.

Like the neuroaffective scientist Jaak Panskepp, Spinoza thought that emotions were the grounding of reason. Emotions can be rational or irrational. It is undoubtedly a triumph to achieve an understanding of the sort we have of the world about us and our place within it and that is due in part to the development of the scientific method one cornerstone of which is to try to achieve objectivity by reducing the levels of emotional attachment you have to anything within the domain you are trying to understand or the prior theories you are seeking to disprove and improve upon. But this is a reduction of the real object and as such a falsification of the wider reality—an abstraction. It is here that I think that the abstraction of the personal self to some entity that might be ported to a machine is likely to find its greatest challenge. One doesn't need to be reminded of the ways in which a minor illness, or appetitive lapse, or the invasion by a virus, or an imbalance in a gland can invade and distort the mind's ability to see the world clearly even when perfectly healthily embodied in a normal human frame.

The development of superintelligent machines should be able to give an account of what superintelligence is and why it is useful, how it differs from general human intelligence or not and what the relationship between reason and emotion is in the natural world and whether that relationship must, or can, or doesn't need to apply in the world of intelligent machines.

As Pascal wrote "The heart has its reasons of which reason knows nothing." There are already well-established areas of research in computer science directed to the area of affective computing. There is also considerable research effort going into neuroaffective and neuropsychoanalytic areas. Psychoanalysis is a bio-psycho-socio science. The importance of the emotions and of your relations to others as a part of the process of coming to be who you are, and your development as an intrinsically social relational being, none of which ever get considered if it is only the narrow cognitive and rational aspects that are being considered. The whole question of the emotions needs to be addressed and understood as part of the wider project of understanding what exactly a superintelligent machine is and how it might function. Kurzweil talks about emotional intelligence, rationality, knowledge and intelligence but rarely if ever about emotion as feelings—rational and irrational. His concerns all lie at the cognitive end of the spectrum of powers that we possess. No little surprise then that he discusses the neo-cortex in particular and its pattern matching capabilities, an area in which his expertise has led to a number of powerful and useful tools. But what if it was emotions and emotional relationships between people that were by far the most important aspect of our apprehension of and activity in the world as Rosalind Picard (2014) in an interview about affective computing suggests when she makes a number of interesting and important points regarding emotion and computing. She comments "Emotion … plays a constant role in our experience … it's always there." And concerning the problem of machines having feelings she argues that "As far as qualia are concerned no one knows how to build that in a machine now … no one can even see how it is possible" (qualia being the subjective or qualitative properties of experience). She goes on to pose significant questions about the future of robotics.

> I don't know if robots will ever have feelings the way that we do … it will be some time before the robots on their own accord go out and seek their rights as robots because they feel unjustly treated and if they do that … it would be because we basically built them for that purpose … we have the choice as the creators of these machines to design them in such a way.

Regarding AI in general and her own developing understanding of the importance of emotion to AI she says, "AI in its first fifty years didn't think emotion was important…" And commenting upon her role in the development of affective computing she says:

> As I learned more about how our brains work … beneath our highly evolved cortex there are sub-cortical structures that are deeply involved with emotion, attention and memory… Emotion is actually key… If we want to build an AI that works in the real world that handles complex unpredictable information with flexibility and intelligence then this is a core part of building an intelligent system.

In a similar vein, Sherry Turkle (2004), who has argued for many years from within the domain of artificial intelligence for a much closer relationship between psychoanalysis and computer science, issues this warning:

> People may be comforted by the notion that we are moving from a psychoanalytic to a computer culture, but what the times demand is a passionate quest for joint citizenship if we are to fully comprehend the human meanings of the new and future objects of our lives. (Turkle 2004, p. 30)

Having previously spelt out in some detail what is going to be required for any such close cooperation between the two disciplines she suggests that

> We must cultivate the richest possible language and methodologies for talking about our increasingly emotional relationships with artifacts. We need far closer examination of how artifacts enter the development of the self and mediate between self and other. Psychoanalysis provides a rich language for distinguishing between need (something that artifacts may have) and desire (which resides in the conjunction of language and flesh). It provides a rich language for exploring the specificity of human meanings in their connections to the body. (ibid. p. 29)

In a number of books and articles Turkle (1978, 1984, 1995, 2011) has drawn attention to both the relationship between psychoanalysis and computing and the increasing threat to human social relations by the ubiquity of human machine relations. Turkle gave a TED talk on the subject of the book in February 2012, under the title "Connected, but alone?" Points from her talk echo those in the book:

> 1. The communication technologies not only change what people do, but also changes who they are. 2. People are developing problems in relating to each other, relating to themselves, and their capacity for self-reflection. 3. People using these devices excessively expect more from technology and less from each other. Technologies are being designed that will give people the illusion of companionship without the demands of friendship. 4. The capacity for being alone is not being cultivated. Being alone seems to be interpreted as an illness that needs to be cured rather than a comfortable state of solitude with many uses. 5. Traditional conversation has given way to mediated connection, leading to the loss of valuable interpersonal skills. (http://en.wikipedia.org/wiki/Sherry_Turkle)

There is a lot of time and money spent on developing so-called caring robots for the treatment of the elderly and the care of children (Independent 2015) when the sensible and obvious solution to the problem of social care is other people, properly trained and properly remunerated, other people. We are by nature social creatures but the social valorization of care, support, attachment, nurture, education etc. has been seriously eroded by a culture that measures things only by their monetary value. For which reason we need to re-frame the whole of the caring process from cradle to grave so that parents, nurses, teachers, social workers, carers etc. are highly valued and highly paid for the work they do in helping to produce a new generation of flourishing adults, for caring for the sick and injured, and for helping to make the last days of sick and ailing adults and children a positive and dignified experience instead of trying to replace a humane and caring environment by cheap and tacky robots. If the development of intelligent machines is going to reduce the necessary workforce for the material reproduction of society then we need to redirect our resources towards the reproduction of our invaluable human resources.

12.5 Psychoanalysis

In *The Singularity is Near* Kurzweil (2009) includes Freud among his imaginary interlocutors but gives him nothing of any interest or import to say. I would like to redress this imbalance by quoting from Freud's *Civilization and its Discontents.*

> Life, as we find it, is too hard for us; it brings us too many pains, disappointments and impossible tasks. In order to bear it we cannot dispense with palliative measures. 'We cannot do without auxiliary constructions', as Theodore Fontane tells us. There are perhaps three such measures: powerful deflections, which cause us to make light of our misery: substitutive satisfactions, which diminish it; and intoxicating substances, which make us insensitive to it. Something of this kind is indispensable. (Freud S.E. 21)

I think that the technological singularity plays such a role for Kurzweil and for many of the other people who believe in it and transhumanism too, and that it deflects our awareness, intoxicates our imaginations and provides substitute satisfactions in place of a real assay of our prospects.

Psychoanalysis is a bio-psycho-social science in my view and it only makes any sense when these three different factors are included. As an object relations thinker I am most interested in the social and psychological aspects of this science but I am also convinced about the necessity of the body for the realisation of these relationships. It isn't just that the body is the obvious site of emotions and the vector of emotional relationships it is also the case that what is unique and interesting about human beings is their specific embodiment within a body, a family, a social milieu, a geographical region, a societal form all with their own histories and dynamics. The downside of this from the point of view of the person living this reality is that people they know and love, to whom they are deeply attached will get sick, have accidents, pick fights, misunderstand them and so on, and die leaving them bereft. But, being embodied too it follows that they will also be riven by emotional concerns and upsets which might well include their own suffering from various illnesses, bad luck, difficulties and so on, leading eventually to death. Psychoanalysis, like religion, is about helping people come to terms with the realities of our lives and relationships and the inevitability of suffering and being able to carry on productively and flourish.

I am going to briefly look at the work of Ronald Fairbairn (1952) who, while strongly influenced by Freud nevertheless criticised the underlying assumptions of Freud's structural model and developed a thoroughgoing object relations theory based upon twentieth century physics where energy and structure are interconnected. His psychology of dynamic structure is a multi agent open system model of the mind that is widely influential and is a direct precursor of attachment theory and relational approaches to psychoanalysis subsequently. The overriding imperative of the neonate in this model is to make and sustain a relationship with the mother or surrogate to ensure its survival, not pleasure seeking as in Freud's thinking (Clarke 2006, 2014).

In Fairbairn's model the original defences are to first internalise a significant other, usually mother, followed by, over time, sorting the relationships with mother, by dissociating those object relationships that are unacceptable because over-exciting or over-rejecting. This is followed by the repression of those sets of object relations possibilities that are either over-exciting or over-rejecting, producing a tripartite division of the mind each partition of which is represented by a subject-object dyad. The acceptable object relations becoming the basis of the conscious central self whilst the other selves—libidinal and antilibidinal—form subsidiary selves comprising the unconscious aspects of inner reality.

One of Fairbairn's first papers in the development of this new approach addressed the idea that we are all fundamentally split to some degree in consequence of our original experience with our first objects. Fairbairn's clinical description is of the schizoid character unable to cope with the emotional world of relationships, who retreats to a 'higher plain' where the intellect alone reigns. This seems to be a common characteristic of many of the people engaged in promoting the technological singularity.

> Another important manifestation of preoccupation with the inner world is the tendency to *intellectualization*; and this is a very characteristic schizoid feature. It constitutes and extremely powerful defensive technique; and it operates as a very formidable resistance in psychoanalytical therapy. Intellectualisation implies an over-valuation of the thought processes; and this over-valuation of thought is related to the difficulty which the individual with a schizoid tendency experiences in making emotional contacts with other people. Owing to preoccupation with the inner world and the repression of affect which follows in its train, he has difficulty in expressing his feelings naturally towards other, and in acting naturally and spontaneously in his relations with them. This leads him to make an effort to work out his emotional problems intellectually in the inner world. (Fairbairn 1952, p. 20)

This is to distance yourself from a direct and emotional engagement with the reality around you for a cleaner and purer place where the world of knowledge exists unencumbered by the messy reality from which it has been extracted.

> The search for intellectual solutions to what are properly emotional problems thus gives rise to two important developments: (1) The thought process becomes highly libidinized; and the world of thought tends to become the predominant sphere of creativity and

> self-expression; and (2) ideas tend to become substituted for feelings, and intellectual values for emotional values. (Fairbairn 1952, p. 20)

As a technique, this stepping back to gain a perspective and then applying the extracted model is an acceptable scientific approach as long as you don't forget that the map is not the territory.

There is however an aspect of science that is, for me, brought out best by the Critical Realism of Bhaskar (1989) in which the very possibility of doing science is predicated upon reality existing independently of our hopes and wishes for it and being able to return results that we didn't expect, which can contradict our hypotheses and our certainties, which is what makes the experimental method so important.

If, as Kurzweil argues, there is no distinction between human and machine or between physical and virtual reality then the possibility for being human will have been lost and the possibility for science equally so. For the virtual to be indistinguishable from the real the level of interrogation that we might subject the virtual to would have to be as multi-layered and eventually unpredictable as the continuing investigation of the real is proving to be. It would also have to come apart in exactly the same way that reality does, but, since it is constructed by us we would have to be able to know how it all fits together beforehand, whereas science leads to our discovering anew how it all fits together. If we already know how it works and can make a virtual world consistent with that then we cannot learn from it.

It seems to me highly likely that what Kurzweil is trying to do by ushering in this new age singularity is to avoid all those painful experiences by hoping for a world without them. A world where our biological bodies will no longer break down and need repair and maintenance, in which old age, sickness and death are effectively banished and in which we will at last come to understand the thinking of perfectly rational others instead of always being perplexed by the thinking and behavior of emotional, opinionated, disputatious human beings. Indeed one could argue that many or indeed most of our most precious achievements as a species are predicated upon our grappling with disappointment, anxiety, perplexity and sadness.

Macmurray (1961) argues that for us to be human we have to be able to make real choices and that the difference between ourselves and other organisms is that we have choices because we have intentionality, that is we can look at a situation and see a number of possible ways that we might behave that are not predictable from a knowledge of our biological makeup and our recent history. This is all predicated upon our not being simple instinct-driven creatures despite Freud.

More to the point perhaps is that the sort of thinking that fuels research towards the aims that the singularity determines is leading to the 'nerds' and the 'bots' ripping-off the economy and the market being rigged by the so-called 'Flash Boys' (Lewis 2014) while the global economy becomes more and more unbalanced with the eighty five richest people as wealthy as the poorest half of the world's population (Wearden 2014).

Singularity considerations never address the question of who is going to be advantaged by it and one has the distinct impression that it is a private fantasy about living forever and avoiding the 'slings and arrows of outrageous fortune' that is the prime motivator for individuals concerned mostly with themselves and their own survival. While there are some signs that there is a growing critical response to these ideas and the hyper-market capitalism they encourage in the work of Lanier (2010, 2013), Lucas (2014) or Piketty (2014) the concentration of power and wealth continues unabated and effectively unopposed to the detriment of all. We cannot do without auxiliary constructions if we are to help redirect the economy towards human ends.

In *Thus Spake Zarathustra* Nietzsche talks about the Ubermensch, the Superman, who he entreats us to make 'the meaning of the earth!' which he immediately follows with another entreaty, to '*remain true to the earth*'. I take it that remaining true to the earth is to remain true to our natural place within the biosphere and to bring all of that to some higher order, some greater harmony. I take it that this is to recognise the importance of our embodiment and the ways in which we come to be ourselves through others. I take it that this is to engage fully, emotionally with others and with our societies to work towards producing communities that value their members flourishing. The technological singularity as currently conceived, which might happen in some form or another if we are not vigilant, will be the triumph of reason over emotion, the worst sort of mechanistic thinking in denial of all of our feelings of community, solidarity, attachment and love. The age of the machines will no longer be human and lacks any unifying project that we might understand. An Artificial General Intelligence (AGI) without consciousness or a conscience that is more like a disease than a panacea and an AGI with consciousness and a conscience, may well attempt to usurp us. Be very careful what you wish for!

12.6 Conclusion

If the gold standard of scientific endeavour is the experimental method and falsifiability then both the singularity and the transhuman hypotheses fail that test. In Sects. 12.2–12.4 I adopt a sceptical view of the many claims that are made by supporters of the singularity and transhuman projects and look at alternative constructions. This seems to me a necessary scientific approach towards these claims many of which are both speculative and from a perspective that is neither proven nor agreed. In this way I am concerned to challenge some of the certainties of different aspects of the current research into AGI's with a view *to generating a far greater awareness of the dangers of this research and encourage its regulation*[1] as we do research into stem cells, germ warfare and atomic weapons.

In Sect. 12.5 having searched the introductory survey to this book in vain for any recognition that a depth psychological approach might be helpful to a full understanding of the phantasies surrounding the development of machines that

[1]This places my approach in the 'Societal Proposals' of the introductory survey.

might offer us the possibility of living forever or achieving immense power intellectually or physically, I outline a psychoanalytic approach to the theories of the singularity and transhumanism. In the survey there are a few references to 'psychology' or 'psychopathy' but none that suggest that an important aspect of the imaginative investment that is being made in these AI related projects may be born of phantasy and none that suggest that we have an unconscious the workings of which are to some degree hidden from us. These phantasies are in the service of defence against all of the ills our embodied selves are heir to—death, old age, illness, accident, the loss of loved ones, the loss of our children. The title of the paper "We cannot do without auxiliary constructions" is a tacit acknowledgement of this need for defensive means that psychoanalysis argues we have and use continually to protect and in some cases delude ourselves. As T.S. Eliot says in Burnt Norton, one of his Four Quartets, "humankind cannot bear very much reality". *The psychoanalytic section of this paper is intended to offer a new thread to further discussion of the societal proposals of the introductory survey.*

The work of philosopher and cultural critic John Gray is a case in point. When in a recent book *The Soul of the Marionette* (2015) Gray argues that Kurzweil's vision of the singularity as "an explosive increase in knowledge that will enable humans to emancipate themselves from the material world and cease to be biological organisms [by] … uploading brain information into cyberspace … [that] The affinities between these ideas and Gnosticism are clear. Here as elsewhere, secular thinking is shaped by *forgotten or repressed* religion." (Emphasis added.)

Perhaps the ultimate irony is that the model that Kurzweil holds most dear in his imagination is that of his 'resurrected' father as a companion. It is clear that he was deeply affected by his father's death and strongly motivated to overcome that in some way. The fact that such attachments and needs are strongly mediated by biological, psychological and social factors seems to have been lost in his understanding of the situation. Computer agents without emotional, relational, social and personal needs and wants are much more likely to be successful in helping us in the just and fair administration of things. And a simulation of another person is always just that however convincing or comforting that might be. Kurzweil (2014) reviewed 'Her' a recent film in which a computer agent with a female persona makes and later breaks a relationship with a man who suffers all the heartache that breaking up with another human intimate can bring. Considering one of the main problems of the relationship—the lack of a body for the computer agent—Kurzweil says "There are also methods to provide the tactile sense that goes along with a virtual body. These will soon be feasible, and will certainly be completely convincing by the time an AI of the level of Samantha is feasible." The actual difficulties of effecting that relationship fully are not broached and the idea that for example one might be installed in a flotation tank while hooked up to an agent that was apparently sharing a fully embodied experience with you seems unlikely to catch on, even if it was realisable. However long you postponed it, you would still have to leave the flotation tank and slough off the second skin, at which time the Latin proverb—*post coitum omne animal triste est*—every animal feels sad after sexual intercourse might apply with the following amendment—every animal feels sad (disappointed, frustrated) after virtual sex too. Blurring the distinctions between the real and the virtual is a perilous path to take in my view.

References

Bhaskar, R. (1989) The Possibility of Naturalism. Routledge: London.

Chomsky, N. The Singularity is Science Fiction http://www.singularityweblog.com/noam-chomsky-the-singularity-is-science-fiction/ (last accessed May 2014).

Clarke, G. S. (2006) Personal Relations Theory: Fairbairn, Macmurray and Suttie. Routledge: London.

Clarke, G. S. and Scharff, D. E. (Eds) (2014) Fairbairn and the object relations tradition. Karnac: London.

Fairbairn, W. R. D. (1952) Psychoanalytic Studies of the Personality. Routledge: London.

Gray, J. (2015) The Soul of the Marionette: A Short Enquiry into Human Freedom, Allen Lane: London.

Hofstadter, D. R. (2007) An interview with Douglas. R. Hofstadter http://www.americanscientist.org/bookshelf/pub/douglas-r-hofstadter (last accessed May 2014).

Hofstadter, D. R. (2008) An interview with Douglas.R.Hofstadter following "I am a strange loop". http://tal.forum2.org/hofstadter_interview (last accessed 5th May 2014).

Hofstadter, D. R. (2009) The Singularity Summit at Stanford. http://www.youtube.com/watch?v=qiepYwKPcxc (last accessed May 2014).

Japanese 'robot with a heart' will care for the elderly and children, (http://www.independent.co.uk/life-style/gadgets-and-tech/japanese-robot-with-a-heart-will-care-for-the-elderly-and-children-9491819.html) (last accessed June 2015).

Kurzweil, R. (2009) The Singularity is Near, Duckworth: London.

Kurzweil, R. (2014) A review of 'Her' by Ray Kurzweil. http://www.kurzweilai.net/a-review-of-her-by-ray-kurzweil (last accessed May 2014).

Lanier, J. (2010) You are not a gadget. Allan Lane: London.

Lanier, J. (2013) Who owns the future. Penguin: London.

Lewis, M. (2014) Flash Boys: A Wall Street Revolt. W.W.Norton and Company.

Lucas, R. (2014) Review of Jaron Lanier "Who Owns the Future". Rob Lucas New Left Review 86 Mar/Apr 2014.

Macmurray, J. (1961) Persons in Relation. Faber and Faber: London.

Modis, T. (2006) The Singularity Myth. Technological Forecasting and Social Change, 72, No 2.

Penrose, R. and Hameroff, S. Orchestrated objective reduction. http://en.wikipedia.org/wiki/Orchestrated_objective_reduction (last accessed June 2015).

Picard, R. blog: http://rdigitalife.com/rosalind-picard-blog/ (last accessed May 2014).

Piketty, T. (2014) Capital in the Twenty-First century. Harvard University Press.

Solms, M. and Panksepp, J. (2012) Abstract: The "Id" knows more than the "Ego" admits: neuropsychoanalytic and primal consciousness perspectives on the interface between affective and cognitive neuroscience. Brain Sciences, 2: 147–175.

Turkle, S. (1978) Psychoanalytic Politics: Jacques Lacan and Freud's French Revolution.

Turkle, S. (1984) The Second Self. MIT Press.

Turkle, S (1995) Life on the screen. Simon and Schuster Paperbacks.

Turkle, S. (2004) Whither psychoanalysis in Computer culture? Psychoanalytic Psychology Vol. 21, No 1. 16–30.

Turkle, S. (2011) Alone Together: Why we expect more from technology and less from each other.

Wearden, G. (2014) Oxfam: 85 richest people as wealthy as poorest half of the world. Guardian, 20th January 2014.

Part III
Reflections on the Journey

Chapter 13
Reflections on the Singularity Journey

James D. Miller

13.1 Introduction

Vernor Vinge's 1993 essay "What is the Singularity?" (reprinted in this book's appendix) seems too compelling to be believable. The author, a science fiction writer and former computer science professor, claims that the likely rise of superintelligence means we are on the "edge of change comparable to the rise of human life on Earth." Vinge proposes several ways we might acquire superintelligence, including through developing sentient computers, large computer networks that "wake up," computer/human interfaces, and through biologically enhanced humans, which seems a far more likely path now than in 1993 due to the rapidly falling cost of gene sequencing and the development of CRISPR gene editing technology. Vinge claims that the Singularity won't necessarily go well for our species, and, as he writes, could result in the "physical extinction of the human race." Alternatively, a positive singularity could create "benevolent gods," with Vinge expressing a desire to become one himself.

If an article explained why the stock value of X Corporation was going to rise slightly next month, I might believe it, but if the article claimed that the stock was going to increase tenfold by the week's end I would reject the argument, regardless of how compelling it seemed. Many others have a far better understanding of the stock market than I do. If the article had merit, these more knowledgeable investors would already have bought up the stock until its value had increased tenfold. Consequently, the fact that other investors have not already acted on the article's arguments provides strong evidence against the article's claims. Analogously, if I, a mere academic economist, find Vinge's arguments compelling and my impression is justified then shouldn't society already have been reshaped to take into account

J.D. Miller (✉)
Department of Economics, Smith College, Northampton, US
e-mail: jdmiller@smith.edu

V. Callaghan et al. (eds.), *The Technological Singularity*,
The Frontiers Collection, DOI 10.1007/978-3-662-54033-6_13

the significant likelihood that a Singularity is near? After all, if we are on the verge of a Singularity that will, with high probability, either exterminate us or bring utopia, and this is something that reasonably informed people should realize, shouldn't achieving a positive Singularity have become the primary purpose of human civilization? And if not, shouldn't technologically literate people be screaming at the rest of mankind that how we handle artificial and enhanced human intelligence over these next few decades is mankind's ultimate test?

13.2 Eliezer Yudkowsky

In one of this section's essays, AI researcher Eliezer Yudkowsky defines the three major schools of the Singularity. Each definition, I believe, comes with a reason why people might have a hard time thinking that the Singularity is plausible.

13.2.1 The Event Horizon

Yudkowsky calls the school that Vinge supports the "Event Horizon." This school is characterized in part by the future being unpredictable since it will be controlled by smarter-than-human intelligences—just as a 5-year-old is incapable of understanding the modern world, we mere unaugmented humans probably can't understand what will happen in a future dominated by superintelligences. But I suspect that since we are a storytelling species the impossibility of imagining the post-Singularity world makes the Singularity seem unbelievable to many. Just as quantum physics feels less believable than Newtonian mechanics because only the latter can be visualized, a Singularity's event horizon, beyond which we cannot see, might cause our brains to unreasonably discount the Singularity's plausibility.

13.2.2 Accelerating Change

Yudkowsky refers to another Singularity school of thought as "Accelerating Change." The most prominent proponent of this school is Ray Kurzweil. The "Accelerating Change" school estimates that we will reach a Singularity through exponential increases in information technology, most notably Moore's law. But humans are linear thinkers, and have great difficulty intuitively understanding exponential growth. Imagine, for example, that you have two investments. The first always pays $1000 a month in dividends and the second pays a sum that doubles each month, but starts out at a very low value. As the second investment will, for a very long time, pay much less than the first, it would be challenging for most people to imagine that if you go far enough into the future the second investment will be the one of overwhelming importance. Similarly, although most people recognize that information technology plays a significant role in our economy, they have a hard time extrapolating to what will happen when this technology is orders of magnitude more powerful than it is today.

13.2.3 The Intelligence Explosion

Yudkowsky calls the school he thinks most likely "The Intelligence Explosion." According to this school an AI has the potential to undergo an intelligence chain reaction in which it quickly upgrades its own intelligence. By this theory, an AI might examine its own code to figure out ways of becoming smarter, but as it succeeds in augmenting its own intelligence it will become even better at figuring out ways of becoming smarter. An AI that undergoes such an intelligence explosion might, therefore, expand from human level intelligence to (what would seem to us) godlike intelligence within a day. Although I personally think Yudkowsky's belief in a future intelligence explosion is reasonable, I admit it seems superficially absurd.

The absurdity heuristic provides another explanation for why many might reject the possibility of a Singularity. Most things that seem absurd are absurd, and are not worth taking the time to study in detail. You shouldn't bother investigating the legitimacy of an email from an alleged Nigerian prince who wants just a bit of your money to unlock his fortune nor should a physics professor devote any of her precious time studying the equations mailed to her purportedly showing how to extract infinite free energy from coconuts. The Singularity, especially through an intelligence explosion, does superficially seem like a crazy idea, one thought up by people spending too much time contemplating the philosophical implications of the "Terminator" movies. Most people who hear about the Singularity might immediately pattern match it to science fiction craziness and banish the concept to the silly ideas trash bin subsystem of their brain. Perhaps it was because Vinge, a science fiction author, enjoyed thinking about futuristic crazy-sounding ideas that he gladly paid the upfront contemplation costs of considering if the Singularity should be taken seriously. This might also explain why, anecdotally, many people in the Singularity community are science fiction fans. Part of the value of the Singularity Hypothesis I and II books is to signal to academic audiences that although the Singularity seems absurd, many thoughtful people who have studied it think that you should ignore the absurdity heuristic and pay the time costs to fully investigate the possibility of a coming Singularity.

13.2.4 MIRI and LessWrong

Yudkowsky helped create the Singularity Institute (now called the Machine Intelligence Research Institute) to help mankind achieve a friendly Singularly. (Disclosure: I have contributed to the Singularity Institute.) Yudkowsky then founded the community blog http://LessWrong.com, which seeks to promote the art of rationality, to raise the sanity waterline, and to in part convince people to make considered, rational charitable donations, some of which, Yudkowsky (correctly) hoped, would go to his organization. LessWrong has had a massive impact on the worldview of people who consider the "Intelligence Explosion" school of the Singularity to be the most plausible. Studying the art of rationality gives us additional reasons why many people don't consider the Singularity to be near.

13.3 Scott Aaronson

The two critical skills an aspiring rationalist needs are, first, admitting that a core belief you hold might be wrong and, second, recognizing that an informed person disagreeing with you should lower your estimate of the probability of your being right. Accepting these hard truths forces this author to admit that the reason why the Singularity seems implausible to so many people might be because it is indeed a highly implausible outcome. In recognition of this possibility this book includes the essay The Singularity is Far by Scott Aaronson, an MIT computer science professor. I find the essay's Copernican argument particularly compelling because if the Singularity is near we live at an extraordinarily special time in the history of mankind: how we handle the Singularity will determine the very long term fate of our species. Imagine you conclude that during your expected lifetime mankind will either go extinct or create a friendly AI that will allow us to survive until the end of the universe with trillions upon trillions of our descendants eventually being born for every human who has ever existed up to today. If we do achieve a positive Singularity then if, at the end of the universe, you ranked in importance everyone who has ever existed, everyone currently alive today would likely be extremely close to the top, probably in the top one trillionth of one percent. We should always be suspicious of theories concluding that we are special—both because most things, tautologically, are not (macro) special, and because of the human bias of thinking that we are more important than is justified by the evidence.

13.4 Stuart Armstrong

Another reason, I suspect, that many discount the Singularity is because we don't have a firm date as to its arrival, and many past predictions of future technological marvels have proven false. It's too easy for people to make far flung predictions (e.g. "in the glorious future X will happen") because either the prediction will come true or you can claim that not enough time has passed for you to have been proved right. Because such indefinite predictions can't be falsified, don't allow their authors' to receive negative feedback, and don't force their authors to risk their reputations, they should be given less weight. For this reason, we have included an essay, originally published on LessWrong, by Stuart Armstrong, an editor of this book, that analyzes 95 time-specific predictions concerning when we will create human-level AI. Armstrong shows that the predictions are not subject to what's called the Maes-Garreau Law, which predicts that predictors will predict that a technology will arrive "just within the lifetime of the predictor!"

13.5 Too Far in the Future

The final reason I wish to discuss as to why many might not be taking the Singularity seriously is because people might be thinking that even if the Singularity is relatively near, there is little about its future nature that we can affect today. Tradeoffs are everywhere, and given that an intelligent allocation of resources can alleviate many problems of the modern world, we do face a cost if we put in effort into contemplating a possible Singularity. Furthermore, as with almost any future problem, if technology continues to advance we will likely have better tools to deal with the Singularity in the future than in the present.

Imagine, though, that in 1900 physicists learned of the possibility of hydrogen bombs, and theorized that by 1970 a total war between powers armed with these weapons could destroy civilization. Would there have been any value to people speculating about a future containing these super-weapons, or would it have been a more productive use of time to wait until the eve of these weapons' development before considering how they would remake warfare? Thinking about the possibility of mutually assured destruction, establishing taboos against using atomic weapons in anger, and designing an international system to stop terrorists and rogue states from acquiring nuclear weapons would, I believe, have been worthwhile endeavor for scholars living in my hypothetical 1900 world. Similarly, as the final essay in this section argues, we should today be studying how to handle advanced AI, even if we are a long time away from being able to code above human level intelligence into our computers.

13.6 Scott Siskind

Scott Siskind, writing under the pseudonym Yvain, was one of LessWrong's most popular writers before he went on to write for his own blog at http://SlateStarCodex.com. Among the many articles he wrote concerning the Singularity is the one included in this section titled, "No Time Like the Present for AI Safety Work." Siskind points out in this essay that there are serious problems related to advanced AI that we can and should work on now.

13.6.1 *Wireheading*

Wireheading occurs when you directly stimulate the pleasure centers of your brain rather than seek happiness by engaging in activities that (indirectly) make you feel better. Siskind points out that wireheading might pose a huge problem for our potential future control over AIs. Simplistically, imagine that an AI's code tells it to maximize X where X is supposed to represent some social welfare objective, but instead of changing the real world the AI just alters its own source code to give X as high a value as its memory allows. And then the AI decides to convert all the atoms it can get a hold of (including those in people) to computer memory so it can further raise X. Wireheading, as Siskind explains, is one of several "very basic problems that affect broad categories of minds." As a result, it's a problem that has a high enough chance of infecting a future AI that working on this issue today has a high expected payoff.

13.6.2 *Work on AI Safety Now*

As Siskind writes, if we achieve a Singularity by an intelligence explosion, then we won't have time to solve wireheading and other AI control problems after we create an AI of near human level intelligence. The best hope for a positive Singularity is if we solve these control problems first. And, I believe, that if our analysis shows that these control problems are intractable, then we have learned something useful as well and we should try to slow down the development of AI while simultaneously accelerating the augmentation of human intelligence so that our species becomes smart enough to figure out how to craft AIs that will actually do X, where X aligns with our true (or at least preferred) values.

Chapter 14
Singularity Blog Insights

James D. Miller

14.1 Three Major Singularity Schools

Eliezer Yudkowsky
Machine Intelligence Research Institute

The following is an edited version of an article that was originally posted on the Machine Intelligence Research Institute blog, September 2007.

Singularity discussions seem to be splitting up into three major schools of thought: Accelerating Change, the Event Horizon, and the Intelligence Explosion.

Accelerating Change

Core claim: Our intuitions about change are linear; we expect roughly as much change as has occurred in the past over our own lifetimes. But technological change feeds on itself, and therefore accelerates. Change today is faster than it was 500 years ago, which in turn is faster than it was 5000 years ago. Our recent past is not a reliable guide to how much change we should expect in the future.

Strong claim: Technological change follows smooth curves, typically exponential. Therefore we can predict with fair precision when new technologies will arrive, and when they will cross key thresholds, like the creation of Artificial Intelligence.

Advocates: Ray Kurzweil, Alvin Toffler(?), John Smart

J.D. Miller (✉)
Department of Economics, Smith College, Northampton, USA
e-mail: jdmiller@smith.edu

V. Callaghan et al. (eds.), *The Technological Singularity*,
The Frontiers Collection, DOI 10.1007/978-3-662-54033-6_14

Event Horizon

Core claim: For the last hundred thousand years, humans have been the smartest intelligences on the planet. All our social and technological progress was produced by human brains. Shortly, technology will advance to the point of improving on human intelligence (brain-computer interfaces, Artificial Intelligence). This will create a future that is weirder by far than most science fiction, a difference-in-kind that goes beyond amazing shiny gadgets.

Strong claim: To know what a superhuman intelligence would do, you would have to be at least that smart yourself. To know where Deep Blue would play in a chess game, you must play at Deep Blue's level. Thus the future after the creation of smarter-than-human intelligence is absolutely unpredictable.

Advocates: Vernor Vinge

Intelligence Explosion

Core claim: Intelligence has always been the source of technology. If technology can significantly improve on human intelligence—create minds smarter than the smartest existing humans—then this closes the loop and creates a positive feedback cycle. What would humans with brain-computer interfaces do with their augmented intelligence? One good bet is that they'd design the next generation of brain-computer interfaces. Intelligence enhancement is a classic tipping point; the smarter you get, the more intelligence you can apply to making yourself even smarter.

Strong claim: This positive feedback cycle goes FOOM, like a chain of nuclear fissions gone critical—each intelligence improvement triggering an average of >1.000 further improvements of similar magnitude—though not necessarily on a smooth exponential pathway. Technological progress drops into the characteristic timescale of transistors (or super-transistors) rather than human neurons. The ascent rapidly surges upward and creates superintelligence (minds orders of magnitude more powerful than human) before it hits physical limits.

Advocates: I. J. Good, Eliezer Yudkowsky

The thing about these three logically distinct schools of Singularity thought is that, while all three core claims support each other, all three strong claims tend to contradict each other.

If you extrapolate our existing version of Moore's Law past the point of smarter-than-human AI to make predictions about 2099, then you are contradicting both the strong version of the Event Horizon (which says you can't make predictions because you're trying to outguess a transhuman mind) and the strong version of the Intelligence Explosion (because progress will run faster once smarter-than-human minds and nanotechnology drop it into the speed phase of transistors).

14.2 AI Timeline Predictions: Are We Getting Better?

Stuart Armstrong
Future of Humanity Institute

The following is an edited version of an article that was originally posted on http://www.LessWrong.com *on August 17, 2012.*

Thanks to some sterling work by Kaj Sotala and others such as Jonathan Wang and Brian Potter—all paid for by the gracious Singularity Institute, we've managed to put together a databases listing all AI predictions that we could find. The list is necessarily incomplete, but we found as much as we could, and collated the data so that we could have an overview of what people have been predicting in the field since Turing.

We retained 257 predictions total, of various quality (in our expanded definition, philosophical arguments such as "computers can't think because they don't have bodies" count as predictions). Of these, 95 could be construed as giving timelines for the creation of human-level AIs. And "construed" is the operative word—very few were in a convenient "By golly, I give a 50% chance that we will have human-level AIs by XXXX" format. Some gave ranges; some were surveys of various experts; some predicted other things (such as child-like AIs, or superintelligent AIs).

Where possible, I collapsed these down to single median estimate, making some somewhat arbitrary choices and judgement calls. When a range was given, I took the mid-point of that range. If a year was given with a 50% likelihood estimate, I took that year. If it was the collection of a variety of expert opinions, I took the prediction of the median expert. If the author predicted some sort of AI by a given date (partial AI or superintelligent AI), I took that date as their estimate rather than trying to correct it in one direction or the other (there were roughly the same number of subhuman AIs as suphuman AIs in the list, and not that many of either). I read extracts of the papers to make judgement calls when interpreting problematic statements like "within 30 years" or "during this century" (is that a range or an end-date?).

So some biases will certainly have crept in during the process. That said, it's still probably the best data we have. So keeping all that in mind, let's have a look at what these guys said (and it was mainly guys).

There are two stereotypes about predictions in AI and similar technologies. The first is the Maes-Garreau law: technologies as supposed to arrive just within the lifetime of the predictor!

The other stereotype is the informal 20–30 year range for any new technology: the predictor knows the technology isn't immediately available, but puts it in a range where people would still be likely to worry about it. And so the predictor gets kudos for addressing the problem or the potential, and is safely retired by the time it (doesn't) come to pass. Are either of these stereotypes born out by the data? Well, here is a histogram of the various "time to AI" predictions.

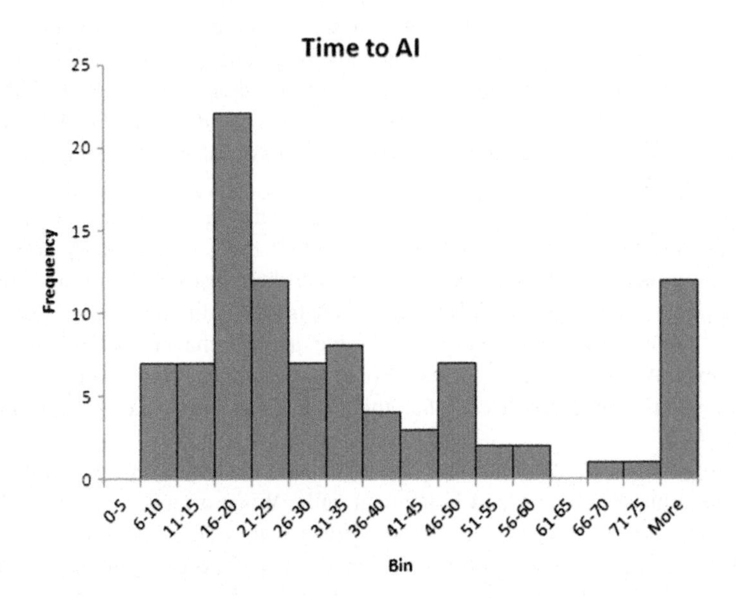

As can be seen, the 20–30 year stereotype is not exactly born out—but a 15–25 one would be. Over a third of predictions are in this range. If we ignore predictions more than 75 years into the future, 40% are in the 15–25 range, and 50% are in the 15–30 range.

Apart from that, there is a gradual tapering off, a slight increase at 50 years, and twelve predictions beyond three quarters of a century. Eyeballing this, there doesn't seem to much evidence for the Maes-Garreau law. Kaj looked into this specifically, plotting (life expectancy) minus (time to AI) versus the age of the predictor; the Maes-Garreau law would expect the data to be clustered around the zero line:

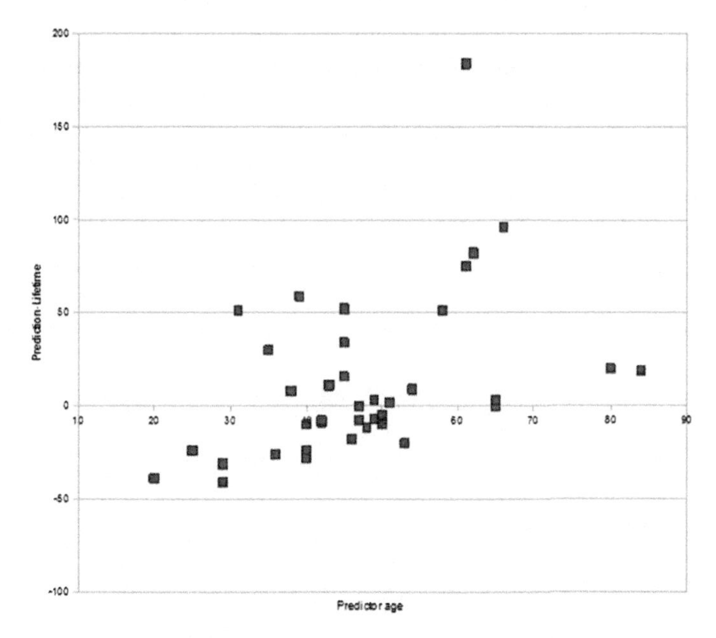

Most of the data seems to be decades out from the zero point (note the scale on the y axis). You could argue, possibly, that 50 year olds are more likely to predict AI just within their lifetime, but this is a very weak effect. I see no evidence for the Maes-Garreau law—of the 37 prediction Kaj retained, only 6 predictions (16%) were within 5 years (in either direction) of the expected death date.

But I've been remiss so far—combining predictions that we know are false (because their deadline has come and gone) with those that could still be true. If we look at predictions that have failed, we get this interesting graph:

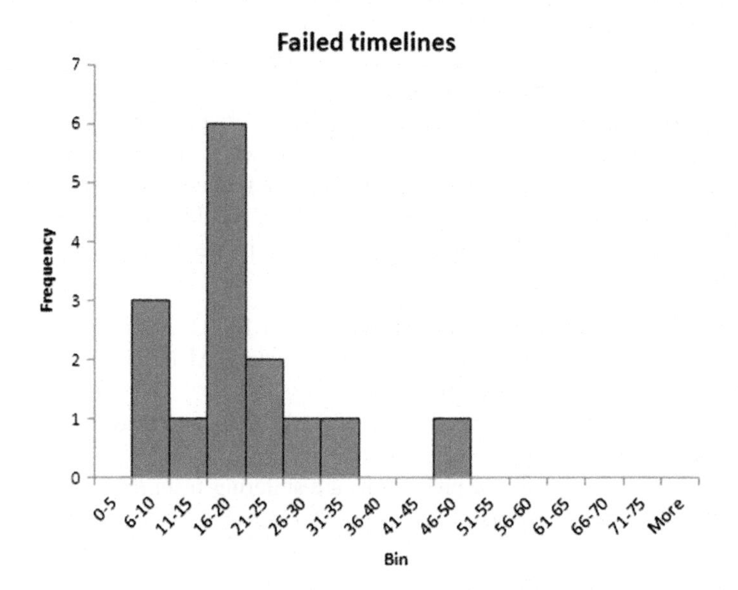

This looks very similar to the original graph. The main difference being the lack of very long range predictions. This is not, in fact, because there has not yet been enough time for these predictions to be proved false, but because prior to the 1990s, there were actually no predictions with a timeline greater than 50 years. This can best be seen on this scatter plot, which plots the time predicted to AI against the date the prediction was made:

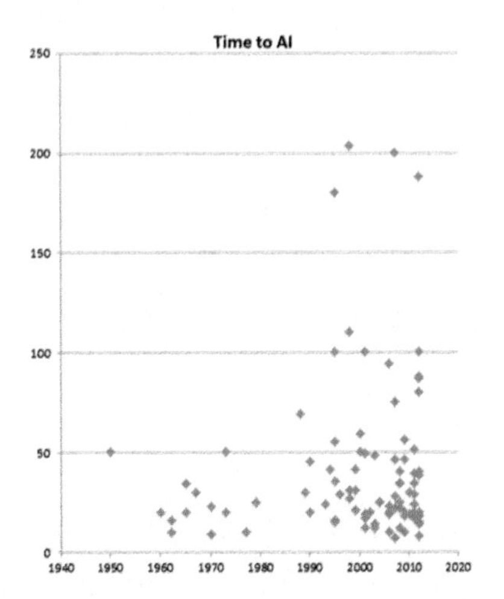

As can be seen, as time elapses, people become more willing to predict very long ranges. But this is something of an artefact—in the early days of computing, people were very willing to predict that AI was impossible. Since this didn't give a timeline, their "predictions" didn't show up on the graph. It recent times, people seem a little less likely to claim AI is impossible, replaced by these "in a century or two" timelines.

Apart from that one difference, predictions look remarkably consistent over the span: modern predictors are claiming about the same time will elapse before AI arrives as their (incorrect) predecessors. This doesn't mean that the modern experts are wrong—maybe AI really is imminent this time round, maybe modern experts have more information and are making more finely calibrated guesses. But in a field like AI prediction, where experts lack feed back for their pronouncements, we should expect them to perform poorly, and for biases to dominate their thinking. This seems the likely hypothesis—it would be extraordinarily unlikely that modern experts, free of biases and full of good information, would reach exactly the same prediction distribution as their biased and incorrect predecessors.

In summary:

- Over a third of predictors claim AI will happen 16–25 years in the future.
- There is no evidence that predictors are predicting AI happening towards the end of their own life expectancy.
- There is little difference between current predictions, and those known to have been wrong previously.
- It is not unlikely that recent predictions are suffering from the same biases and errors as their predecessors.

14.3 No Time Like the Present for AI Safety Work

By Scott Siskind

The following is an edited version of an article that was originally posted on http://www.SlateStarCodex.com on May 29, 2015.

Why we should take AI risk seriously:

1. If humanity doesn't blow itself up, eventually we will create human-level AI.
2. If humanity creates human-level AI, technological progress will continue and eventually reach far-above-human-level AI
3. If far-above-human-level AI comes into existence, eventually it will so overpower humanity that our existence will depend on its goals being aligned with ours
4. It is possible to do useful research now which will improve our chances of getting the AI goal alignment problem right
5. Given that we can start research now we probably should, since leaving it until there is a clear and present need for it is unwise

I place very high confidence (>95%) on each of the first three statements—they're just saying that if trends continue moving towards a certain direction without stopping, eventually they'll get there. I have lower confidence (around 50%) on the last two statements.

Dealing with AI risk is as important as all the other things we consider important, like curing diseases and detecting asteroids and saving the environment. That requires at least a little argument for why progress should indeed be possible at this early stage.

And I think progress is possible insofar as this is a philosophical and not a technical problem. Right now the goal isn't "write the code that will control the future AI", it's "figure out the broad category of problem we have to deal with." Let me give two examples of open problems to segue into a discussion of why these problems are worth working on now.

II.

Problem 1: Wireheading

Some people have gotten electrodes implanted in their brains for therapeutic or research purposes. When the electrodes are in certain regions, most notably the lateral hypothalamus, the people become obsessed with stimulating them as much as possible. If you give them the stimulation button, they'll press it thousands of times per hour; if you try to take the stimulation button away from them, they'll defend it with desperation and ferocity. Their life and focus narrows to a pinpoint, normal goals like love and money and fame and friendship forgotten in the relentless drive to stimulate the electrode as much as possible.

This fits pretty well with what we know of neuroscience. The brain (OVER-SIMPLIFICATION WARNING) represents reward as electrical voltage at a couple of reward centers, then does whatever tends to maximize that reward. Normally this works pretty well; when you fulfill a biological drive like food or sex, the reward center responds with little bursts of reinforcement, and so you continue fulfilling your biological drives. But stimulating the reward center directly with an electrode increases it much more than waiting for your brain to send little bursts of stimulation the natural way, so this activity is by definition the most rewarding possible. A person presented with the opportunity to stimulate the reward center directly will forget about all those indirect ways of getting reward like "living a happy life" and just press the button attached to the electrode as much as possible.

This doesn't even require any brain surgery—drugs like cocaine and meth are addictive in part because they interfere with biochemistry to increase the level of stimulation in reward centers.

And computers can run into the same issue. I can't find the link, but I do remember hearing about an evolutionary algorithm designed to write code for some application. It generated code semi-randomly, ran it by a "fitness function" that assessed whether it was any good, and the best pieces of code were "bred" with each other, then mutated slightly, until the result was considered adequate.

They ended up, of course, with code that hacked the fitness function and set it to some absurdly high integer.

These aren't isolated incidents. Any mind that runs off of reinforcement learning with a reward function—and this seems near-universal in biological life-forms and is increasingly common in AI—will have the same design flaw. The main defense against it this far is simple lack of capability: most computer programs aren't smart enough for "hack your own reward function" to be an option; as for humans, our reward centers are hidden way inside our heads where we can't get to it. A hypothetical superintelligence won't have this problem: it will know exactly where its reward center is and be intelligent enough to reach it and reprogram it.

The end result, unless very deliberate steps are taken to prevent it, is that an AI designed to cure cancer hacks its own module determining how much cancer has been cured and sets it to the highest number its memory is capable of representing. Then it goes about acquiring more memory so it can represent higher numbers. If it's superintelligent, its options for acquiring new memory include "take over all the computing power in the world" and "convert things that aren't computers into computers." Human civilization is a thing that isn't a computer.

This is not some exotic failure mode that a couple of extremely bizarre designs can fall into; this may be the natural course for a sufficiently intelligent reinforcement learner.

Problem 2: The Evil Genie Effect

Everyone knows the problem with computers is that they do what you say rather than what you mean. Nowadays that just means that a program runs differently when you forget a close-parenthesis, or websites show up weird if you put the HTML codes in the wrong order. But it might lead an artificial intelligence to seriously misinterpret natural language orders.

Age of Ultron actually gets this one sort of right. Tony Stark orders his super-robot Ultron to bring peace to the world; Ultron calculates that the fastest and most certain way to bring peace is to destroy all life. As far as I can tell, Ultron is totally 100% correct about this and in some real-world equivalent that is exactly what would happen. We would get pretty much the same effect by telling an AI to "cure cancer" or "end world hunger" or any of a thousand other things.

Even Isaac Asimov's Three Laws of Robotics would take about thirty seconds to become horrible abominations. The First Laws says a robot cannot harm a human being or allow through inaction a human being to come to harm. "Not taking over the government and banning cigarettes" counts as allowing through inaction a human being to come to harm. So does "not locking every human in perfectly safe stasis fields for all eternity."

There is no way to compose an order specific enough to explain exactly what we mean by "do not allow through inaction a human to come to harm"—go ahead, try it—unless the robot is already willing to do what we mean, rather than what we say. This is not a deal-breaker, since AIs may indeed be smart enough to understand what we mean, but our desire that they do so will have to be programmed into them directly, from the ground up.

But this just leads to a second problem: we don't always know what we mean by something. The question of "how do we balance the ethical injunction to keep people safe with the ethical injunction to preserve human freedom?" is a pretty hot topic in politics right now, presenting itself in everything from gun control to banning Big Gulp cups. It seems to involve balancing out everything we value—how important are Big Gulp cups to us, anyway?—and combining cost-benefit calculations with sacred principles. Any AI that couldn't navigate that moral labyrinth might end up ending world hunger by killing all starving people, or refusing else to end world hunger by inventing new crops because the pesticides for them might kill an insect.

This is a problem we have yet to solve with humans—most of the humans in the world have values that we consider abhorrent, and accept tradeoffs we consider losing propositions. Dealing with an AI whose mind is no more different to mine than that of fellow human being Pat Robertson would from my perspective be a clear-cut case of failure.

III.

My point in raising these two examples wasn't to dazzle anybody with interesting philosophical issues. It's to prove a couple of points:

First, there are some very basic problems that affect broad categories of minds, like "all reinforcement learners" or "all minds that make decisions with formal math". People often speculate that at this early stage we can't know anything about the design of future AIs. But I would find it extraordinarily surprising if they used neither reinforcement learning or formal mathematical decision-making.

Second, these problems aren't obvious to most people. These are weird philosophical quandaries, not things that are obvious to everybody with even a little bit of domain knowledge.

Third, these problems have in fact been thought of. Somebody, whether it was a philosopher or a mathematician or a neuroscientist, sat down and thought "Hey, wait, reinforcement learners are naturally vulnerable to wireheading, which would explain why this same behavior shows up in all of these different domains."

Fourth, these problems suggest research programs that can be pursued right now, at least in a preliminary way. How come a human can understand the concept of wireheading, yet not feel any compulsion to seek a brain electrode to wirehead themselves with? Is there a way to design a mind that could wirehead a few times, feel and understand the exact sensation, and yet feel no compulsion to wirehead further? How could we create an idea of human ethics and priorities formal enough to stick into a computer?

I think when people hear "we should start, right now in 2015, working on AI goal alignment issues" they think that somebody wants to write a program that can be imported directly into a 2075 AI to provide it with an artificial conscience. Then they think "No way you can do something that difficult this early on."

But that isn't what anybody's proposing. What we're proposing is to get ourselves acquainted with the general philosophical problems that affect a broad subset of minds, then pursue the neuroscientific, mathematical, and philosophical investigations necessary to have a good understanding of them by the time the engineering problem comes up.

IV.

That last section discussed my claim 4, that there's research we can do now that will help. That leaves claim 5—given that we can do research now, we should, because we can't just trust our descendents in the crunch time to sort things out on their own without our help, using their better model of what eventual AI might look like. There are a couple of reasons for this

Reason 1: The Treacherous Turn

Our descendents' better models of AI might be actively misleading. Things that work for subhuman or human level intelligences might fail for superhuman intelligences. Empirical testing won't be able to figure this out without help from armchair philosophy.

Pity poor evolution. It had hundreds of millions of years to evolve defenses against heroin—which by the way affects rats much as it does humans—but it never bothered. Why not? Because until the past century, there wasn't anything around intelligent enough to synthesize pure heroin. So heroin addiction just wasn't a problem anything had to evolve to deal with. A brain design that looks pretty good in stupid animals like rats and cows becomes very dangerous when put in the hands (well, heads) of humans smart enough to synthesize heroin or wirehead their own pleasure centers.

The same is true of AI. Dog-level AIs aren't going to learn to hack their own reward mechanism. Even human level AIs might not be able to—I couldn't hack a robot reward mechanism if it were presented to me. Superintelligences can. What we might see is reinforcement-learning AIs that work very well at the dog level, very well at the human level, then suddenly blow up at the superhuman level, by which it's time it's too late to stop them.

This is a common feature of AI safety failure modes. If you tell me, as a mere human being, to "make peace", then my best bet might be to become Secretary-General of the United Nations and learn to negotiate very well. Arm me with a few thousand nukes, and it's a different story. A human-level AI might pursue its peace-making or cancer-curing or not-allowing-human-harm-through-inaction-ing through the same prosocial avenues as humans, then suddenly change once it became superintelligent and new options became open. Indeed, the point that will activate the shift is precisely that no humans are able to stop it. If humans can easily shut an AI down, then the most effective means of curing cancer will be for it to research new medicines (which humans will support); if humans can no longer stop an AI, the most effective means of curing cancer is destroying humanity (since it will no longer matter that humans will fight back).

Reason 2: Hard Takeoff

It seems in theory that by hooking a human-level AI to a calculator app, we can get it to the level of a human with lightning-fast calculation abilities. By hooking it up to Wikipedia, we can give it all human knowledge. By hooking it up to a couple extra gigabytes of storage, we can give it photographic memory. By giving it a few more processors, we can make it run a hundred times faster, such that a problem that takes a normal human a whole day to solve only takes the human-level AI 15 min.

So we've already gone from "mere human intelligence" to "human with all knowledge, photographic memory, lightning calculations, and solves problems a hundred times faster than anyone else." This suggests that "merely human level intelligence" isn't mere.

The next problem is "recursive self-improvement". Maybe this human-level AI armed with photographic memory and a hundred-time-speedup takes up computer science. Maybe, with its ability to import entire textbooks in seconds, it becomes very good at computer science. This would allow it to fix its own algorithms to make itself even more intelligent, which would allow it to see new ways to make itself even more intelligent, and so on. The end result is that it either reaches some natural plateau or becomes superintelligent in the blink of an eye.

If it's the second one, "wait for the first human-level intelligences and then test them exhaustively" isn't going to cut it. The first human-level intelligence will become the first superintelligence too quickly to solve even the first of the hundreds of problems involved in machine goal-alignment.

And although I haven't seen anyone else bring this up, I'd argue that even the hard-takeoff scenario might be underestimating the risks.

Imagine that for some reason having two hundred eyes is the killer app for evolution. A hundred ninety-nine eyes are useless, no better than the usual two, but once you get two hundred, your species dominates the world forever.

The really hard part of having two hundred eyes is evolving the eye at all. After you've done that, having two hundred of them is very easy. But it might be that it would take eons and eons before any organism reached the two hundred eye sweet spot. Having dozens of eyes is such a useless waste of energy that evolution might never get to the point where it could test the two-hundred-eyed design.

Consider that the same might be true for intelligence. The hard part is evolving so much as a tiny rat brain. Once you've got that, getting a human brain, with its world-dominating capabilities, is just a matter of scaling up. But since brains are metabolically wasteful and not that useful before the technology-discovering point, it took eons before evolution got there.

There's a lot of evidence that this is true. First of all, humans evolved from chimps in just a couple of million years. That's too short to redesign the mind from the ground up, or even invent any interesting new evolutionary "technologies". It's just enough time for evolution to alter the scale and add a couple of efficiency tweaks. But monkeys and apes were around for tens of millions of years before evolution bothered.

Second, dolphins are almost as intelligent as humans. But they last shared a common ancestor with us something like fifty million years ago. Either humans and dolphins both evolved fifty million years worth of intelligence "technologies" independently of each other, or else the most recent common ancestor had most of what was necessary for intelligence and humans and dolphins were just the two animals in that vast family tree for whom using them to their full extent became useful. But the most recent common ancestor of humans and dolphins was probably not much more intelligent than a rat itself.

If this is right, then the first rat-level AI will contain most of the interesting discoveries needed to build the first human-level AI and the first superintelligent AI. People tend to say things like "Well, we might have AI as smart as a rat soon, but it will be a long time after that before they're anywhere near human-level". But that's assuming you can't turn the rat into the human just by adding more processing power or more simulated neurons or more connections or whatever. Anything done on a computer doesn't need to worry about metabolic restrictions.

Reason 3: Everyday Ordinary Time Constraints

During the 1956 Dartmouth Conference on AI, top researchers made a plan toward reaching human-level artificial intelligence, and gave themselves 2 months to teach computers to understand human language. In retrospect, this might have been mildly optimistic.

But now machine translation is a thing, people are making some good progress in some of the hard problems—and when people bring up problems like wire-heading, or goal alignment, people just say "Oh, we have plenty of time".

But expecting to solve those problems in a few years might be just as optimistic as expecting to solve machine language translation in 2 months. Sometimes problems are harder than you think, and it's worth starting on them early just in case.

14.4 The Singularity Is Far

By Scott Aaronson
The University of Texas at Austin

The following is an edited version of an article that was originally posted on http://www.scottaaronson.com *on September 7, 2008.*

In this post, I wish to propose for the reader's favorable consideration a doctrine that will strike many in the nerd community as strange, bizarre, and paradoxical, but that I hope will at least be given a hearing. The doctrine in question is this: while it is possible that, a century hence, humans will have built molecular nanobots and superintelligent AIs, uploaded their brains to computers, and achieved eternal life, these possibilities are not quite so likely as commonly supposed, nor do they obviate the need to address mundane matters such as war, poverty, disease, climate change, and helping Democrats win elections.

Last week I read Ray Kurzweil's *The Singularity Is Near*, which argues that by 2045, or somewhere around then, advances in AI, neuroscience, nanotechnology, and other fields will let us transcend biology, upload our brains to computers, and achieve the dreams of the ancient religions, including eternal life and whatever simulated sex partners we want. Perhaps surprisingly, Kurzweil does not come across as a wild-eyed fanatic, but as a humane idealist; the text is thought-provoking and occasionally even wise.

I find myself in agreement with Kurzweil on three fundamental points. Firstly, that whatever purifying or ennobling qualities suffering might have, those qualities are outweighed by suffering's fundamental suckiness. If I could press a button to free the world from loneliness, disease, and death—the downside being that life might become banal without the grace of tragedy—I'd probably hesitate for about five seconds before lunging for it.

Secondly, there's nothing bad about overcoming nature through technology. Humans have been in that business for at least 10,000 years. Now, it's true that fanatical devotion to particular technologies—such as the internal combustion engine—might well cause the collapse of human civilization and the permanent degradation of life on Earth. But the only plausible solution is better technology, not the Flintstone route.

Thirdly, were there machines that pressed for recognition of their rights with originality, humour, and wit, we'd have to give it to them. And if those machines quickly rendered humans obsolete, I for one would salute our new overlords. Yet while I share Kurzweil's ethical sense, I don't share his technological optimism. Everywhere he looks, Kurzweil sees Moore's-Law-type exponential trajectories—not just for transistor density, but for bits of information, economic output, the resolution of brain imaging, the number of cell phones and Internet hosts, the cost of DNA sequencing... you name it, he'll plot it on a log scale. Kurzweil acknowledges that, even over the brief periods that his exponential curves cover, they have hit occasional snags, like (say) the Great Depression or World War II. And he's not so naïve as to extend the curves indefinitely. Nevertheless, he fully expects current technological trends to continue pretty much unabated until they hit fundamental physical limits.

I'm much less sanguine. Where Kurzweil sees a steady march of progress interrupted by occasional hiccups, I see a few fragile and improbable victories against a backdrop of malice, stupidity, and greed—the tiny amount of good humans have accomplished in constant danger of drowning in a sea of blood and tears, as happened to so many of the civilizations of antiquity. The difference is that this time, human idiocy is playing itself out on a planetary scale; this time we can finally ensure that there are no survivors left to start over.

In the rest of this post, I'd like to share some of the reasons why I haven't chosen to spend my life worrying about the Singularity. The first, and most important, reason is because there are vastly easier prerequisite questions that we already don't know how to answer. In a field like computer science theory, you very quickly get used to being able to state a problem with perfect clarity, knowing exactly what would constitute a solution, and still not having any clue how to solve it. And at least in my experience, being pounded with this situation again and again slowly reorients your worldview. You learn to terminate trains of thought that might otherwise run forever without halting. Faced with a question like "How can we stop death?" or "How can we build a human-level AI?" you learn to respond: "What's another question that's easier to answer, and that probably has to be answered anyway before we have any chance on the original one?" And if someone says, "but can't you at least estimate how long it will take to answer the original question?" you learn to hedge and equivocate.

The second reason is that as a goal recedes to infinity, the probability increases that as we approach it, we'll discover some completely unanticipated reason why it wasn't the right goal anyway. You might ask: what is it that we could possibly learn about neuroscience, biology, or physics, that would make us slap our foreheads and realize that uploading our brains to computers was a harebrained idea from the start, reflecting little more than early-21st-century prejudice? Is there any example of a prognostication about the 21st century written before 1950, most of which doesn't now seem quaint?

The third reason is simple comparative advantage. Given our current ignorance, there seems to me to be relatively little worth saying about the Singularity—and what is worth saying is already being said well by others. Thus, I find nothing wrong with a few people devoting their lives to Singulatarianism, just as others should arguably spend their lives worrying about asteroid collisions. But precisely because smart people do devote brain-cycles to these possibilities, the rest of us have correspondingly less need to.

The fourth reason is because I find it unlikely that we're extremely special. Sure, maybe we're at the very beginning of the human story, a mere awkward adolescence before billions of glorious post-Singularity years ahead. But whatever intuitions cause us to expect that could easily be leading us astray. Suppose that all over the universe, civilizations arise and continue growing exponentially until they exhaust their planets' resources and kill themselves out. In that case, almost every conscious being brought into existence would find itself extremely close to its civilization's death throes. If—as many believe—we're quickly approaching the earth's carrying capacity, then we'd have not the slightest reason to be surprised by that apparent coincidence. To be human would, in the vast majority of cases, mean to be born into a world of air travel and Burger King and imminent global catastrophe. It would be like some horrific Twilight Zone episode, with all the joys and labors, the triumphs and setbacks of developing civilizations across the universe receding into demographic insignificance next to their final, agonizing howls of pain. I wish reading the news every morning furnished me with more reasons not to be haunted by this vision of existence.

The fifth reason is my (limited) experience of AI research. I was actually an AI person long before I became a theorist. When I was 12, I set myself the modest goal of writing a BASIC program that would pass the Turing Test by learning from experience and following Asimov's Three Laws of Robotics. I coded up a really nice tokenizer and user interface, and only got stuck on the subroutine that was supposed to understand the user's question and output an intelligent, Three-Laws-obeying response. Later, at Cornell, I was lucky to learn from Bart Selman, and worked as an AI programmer for Cornell's RoboCupteam—an experience that taught me little about the nature of intelligence but a great deal about how to make robots pass a ball. At Berkeley, my initial focus was on machine learning and statistical inference; had it not been for quantum computing, I'd probably still be doing AI today. For whatever it's worth, my impression was of a field with plenty of exciting progress, but which has (to put it mildly) some ways to go before recapitulating the last billion years of evolution. The idea that a field must either be (1) failing or (2) on track to reach its ultimate goal within our lifetimes, seems utterly without support in the history of science (if understandable from the standpoint of both critics and enthusiastic supporters). If I were forced at gunpoint to guess, I'd say that human-level AI seemed to me like a slog of many more centuries or millennia (with the obvious potential for black swans along the way).

As you may have gathered, I don't find the Singulatarian religion so silly as not to merit a response. Not only is the "Rapture of the Nerds" compatible with all known laws of physics; if humans survive long enough it might even come to pass. The one notion I have real trouble with is that the AI-beings of the future would be no more comprehensible to us than we are to dogs (or mice, or fish, or snails). After all, we might similarly expect that there should be models of computation as far beyond Turing machines as Turing machines are beyond finite automata. But in the latter case, we know the intuition is mistaken. There is a ceiling to computational expressive power. Get up to a certain threshold, and every machine can simulate every other one, albeit some slower and others faster. Now, it's clear that a human who thought at ten thousand times our clock rate would be a pretty impressive fellow. But if that's what we're talking about, then we don't mean a point beyond which history completely transcends us, but "merely" a point beyond which we could only understand history by playing it in extreme slow motion.

Yet while I believe the latter kind of singularity is possible, I'm not at all convinced of Kurzweil's thesis that it's "near" (where "near" means before 2045, or even 2300). I see a world that really did change dramatically over the last century, but where progress on many fronts (like transportation and energy) seems to have slowed down rather than sped up; a world quickly approaching its carrying capacity, exhausting its natural resources, ruining its oceans, and supercharging its climate; a world where technology is often powerless to solve the most basic problems, millions continue to die for trivial reasons, and democracy isn't even clearly winning over despotism; a world that finally has a communications network with a decent search engine but that still hasn't emerged from the tribalism and ignorance of the Pleistocene. And I can't help thinking that, before we transcend the human condition and upload our brains to computers, a reasonable first step might be to bring the 18th-century Enlightenment to the 98% of the world that still hasn't gotten the message.

Appendix

The Coming Technological Singularity: How to Survive in the Post-human Era (reprint)

Vernor Vinge
Department of Mathematical Sciences, San Diego State University

 This article was for the VISION-21 Symposium sponsored by NASA Lewis Research Center and the Ohio Aerospace Institute, March 30–31, 1993. It is also retrievable from the NASA technical reports server as part of NASA CP-10129. A slightly changed version appeared in the Winter 1993 issue of Whole Earth Review.

Abstract

Within thirty years, we will have the technological means to create superhuman intelligence. Shortly after, the human era will be ended.

Is such progress avoidable? If not to be avoided, can events be guided so that we may survive? These questions are investigated. Some possible answers (and some further dangers) are presented.

What is The Singularity?

The acceleration of technological progress has been the central feature of this century. I argue in this paper that we are on the edge of change comparable to the rise of human life on Earth. The precise cause of this change is the imminent creation by technology of entities with greater than human intelligence. There are several means by which science may achieve this breakthrough (and this is another reason for having confidence that the event will occur):

- The development of computers that are "awake" and superhumanly intelligent. (To date, most controversy in the area of AI relates to whether we can create human equivalence in a machine. But if the answer is "yes, we can", then there is little doubt that beings more intelligent can be constructed shortly thereafter.

V. Callaghan et al. (eds.), *The Technological Singularity*,
The Frontiers Collection, DOI 10.1007/978-3-662-54033-6

- Large computer networks (and their associated users) may "wake up" as a superhumanly intelligent entity.
- Computer/human interfaces may become so intimate that users may reasonably be considered superhumanly intelligent.
- Biological science may find ways to improve upon the natural human intellect.

The first three possibilities depend in large part on improvements in computer hardware. Progress in computer hardware has followed an amazingly steady curve in the last few decades (Moravec 1988). Based largely on this trend, I believe that the creation of greater than human intelligence will occur during the next thirty years. (Platt 1997) has pointed out the AI enthusiasts have been making claims like this for the last thirty years. Just so I'm not guilty of a relative-time ambiguity, let me more specific: I'll be surprised if this event occurs before 2005 or after 2030.)

What are the consequences of this event? When greater-than-human intelligence drives progress, that progress will be much more rapid. In fact, there seems no reason why progress itself would not involve the creation of still more intelligent entities—on a still-shorter time scale. The best analogy that I see is with the evolutionary past: Animals can adapt to problems and make inventions, but often no faster than natural selection can do its work—the world acts as its own simulator in the case of natural selection. We humans have the ability to internalize the world and conduct "what if's" in our heads; we can solve many problems thousands of times faster than natural selection. Now, by creating the means to execute those simulations at much higher speeds, we are entering a regime as radically different from our human past as we humans are from the lower animals.

From the human point of view this change will be a throwing away of all the previous rules, perhaps in the blink of an eye, an exponential runaway beyond any hope of control. Developments that before were thought might only happen in "a million years" (if ever) will likely happen in the next century. (In Bear (1983), Greg Bear paints a picture of the major changes happening in a matter of hours.)

I think it's fair to call this event a singularity ("the Singularity" for the purposes of this paper). It is a point where our models must be discarded and a new reality rules. As we move closer and closer to this point, it will loom vaster and vaster over human affairs till the notion becomes a commonplace. Yet when it finally happens it may still be a great surprise and a greater unknown. In the 1950s there were very few who saw it: Ulam (1958) paraphrased John von Neumann as saying:

> *One conversation centered on the ever accelerating progress of technology and changes in the mode of human life, which gives the appearance of approaching some essential singularity in the history of the race beyond which human affairs, as we know them, could not continue.*

Von Neumann even uses the term singularity, though it appears he is still thinking of normal progress, not the creation of superhuman intellect. (For me, the superhumanity is the essence of the Singularity. Without that we would get a glut of technical riches, never properly absorbed (see Stent 1969)).

In the 1960s there was recognition of some of the implications of superhuman intelligence. Good wrote (1980):

Let an ultraintelligent machine be defined as a machine that can far surpass all the intellectual activities of any any man however clever. Since the design of machines is one of these intellectual activities, an ultraintelligent machine could design even better machines; there would then unquestionably be an "intelligence explosion," and the intelligence of man would be left far behind. Thus the first ultraintelligent machine is the _last_ invention that man need ever make, provided that the machine is docile enough to tell us how to keep it under control.

. . .

It is more probable than not that, within the twentieth century, an ultraintelligent machine will be built and that it will be the last invention that man need make.

Good has captured the essence of the runaway, but does not pursue its most disturbing consequences. Any intelligent machine of the sort he describes would not be humankind's "tool"—any more than humans are the tools of rabbits or robins or chimpanzees.

Through the '60s and '70s and '80s, recognition of the cataclysm spread (Vinge 1987; Alfvén 1969; Vinge1983; Bear 1985). Perhaps it was the science-fiction writers who felt the first concrete impact. After all, the "hard" science-fiction writers are the ones who try to write specific stories about all that technology may do for us. More and more, these writers felt an opaque wall across the future. Once, they could put such fantasies millions of years in the future (Stapledon 1937). Now they saw that their most diligent extrapolations resulted in the unknowable ... soon. Once, galactic empires might have seemed a Post-Human domain. Now, sadly, even interplanetary ones are.

What about the '90s and the '00s and the '10s, as we slide toward the edge? How will the approach of the Singularity spread across the human world view? For a while yet, the general critics of machine sapience will have good press. After all, till we have hardware as powerful as a human brain it is probably foolish to think we'll be able to create human equivalent (or greater) intelligence. (There is the far-fetched possibility that we could make a human equivalent out of less powerful hardware, if were willing to give up speed, if we were willing to settle for an artificial being who was literally slow (Vinge 1987b). But it's much more likely that devising the software will be a tricky process, involving lots of false starts and experimentation. If so, then the arrival of self-aware machines will not happen till after the development of hardware that is substantially more powerful than humans' natural equipment.)

But as time passes, we should see more symptoms. The dilemma felt by science fiction writers will be perceived in other creative endeavors. (I have heard thoughtful comic book writers worry about how to have spectacular effects when everything visible can be produced by the technically commonplace.) We will see automation replacing higher and higher level jobs. We have tools right now (symbolic math programs, cad/cam) that release us from most low-level drudgery. Or put another way: The work that is truly productive is the domain of a steadily smaller and more elite fraction of humanity. In the coming of the Singularity, we are seeing the predictions of true technological unemployment finally come true.

Another symptom of progress toward the Singularity: ideas themselves should spread ever faster, and even the most radical will quickly become commonplace. When I began writing, it seemed very easy to come up with ideas that took decades to percolate into the cultural consciousness; now the lead time seems more like eighteen months. (Of course, this could just be me losing my imagination as I get old, but I see the effect in others too.) Like the shock in a compressible flow, the Singularity moves closer as we accelerate through the critical speed.

And what of the arrival of the Singularity itself? What can be said of its actual appearance? Since it involves an intellectual runaway, it will probably occur faster than any technical revolution seen so far. The precipitating event will likely be unexpected—perhaps even to the researchers involved. ("But all our previous models were catatonic! We were just tweaking some parameters....") If networking is widespread enough (into ubiquitous embedded systems), it may seem as if our artifacts as a whole had suddenly wakened.

And what happens a month or two (or a day or two) after that? I have only analogies to point to: The rise of humankind. We will be in the Post-Human era. And for all my rampant technological optimism, sometimes I think I'd be more comfortable if I were regarding these transcendental events from one thousand years remove ... instead of twenty.

Can the Singularity be Avoided?

Well, maybe it won't happen at all: Sometimes I try to imagine the symptoms that we should expect to see if the Singularity is not to develop. There are the widely respected arguments of Penrose (1989) and Searle (1980) against the practicality of machine sapience. In August of 1992, Thinking Machines Corporation held a workshop to investigate the question "How We Will Build a Machine that Thinks" [Thearling]. As you might guess from the workshop's title, the participants were not especially supportive of the arguments against machine intelligence. In fact, there was general agreement that minds can exist on nonbiological substrates and that algorithms are of central importance to the existence of minds. However, there was much debate about the raw hardware power that is present in organic brains. A minority felt that the largest 1992 computers were within three orders of magnitude of the power of the human brain. The majority of the participants agreed with Moravec's estimate (Moravec 1988) that we are ten to forty years away from hardware parity. And yet there was another minority who pointed to Conrad et al. (1989), Rasmussen et al. (1991), and conjectured that the computational competence of single neurons may be far higher than generally believed. If so, our present computer hardware might be as much as _ten_ orders of magnitude short of the equipment we carry around in our heads. If this is true (or for that matter, if the Penrose or Searle critique is valid), we might never see a Singularity. Instead, in the early '00s we would find our hardware performance curves begin to level off—this caused by our inability to automate the complexity of the design work necessary to support the hardware trend curves. We'd end up with some very powerful hardware, but without the ability to push it further. Commercial digital signal processing might be awesome, giving an analog appearance even to digital operations, but nothing would ever "wake up" and there would never be the intellectual runaway which is the essence of the Singularity. It would likely be seen as a golden age ... and it would also be an end of progress. This is very like the future predicted by Gunther Stent. In fact, on page 137 of Stent (1969), Stent explicitly cites the development of transhuman intelligence as a sufficient condition to break his projections.

But if the technological Singularity can happen, it will. Even if all the governments of the world were to understand the "threat" and be in deadly fear of it, progress toward the goal would continue. In fiction, there have been stories of laws passed forbidding the construction of "a machine in the form of the mind of man" (Herbert 1985). In fact, the competitive advantage—economic, military, even artistic—of every advance in automation is so compelling that passing laws, or having customs, that forbid such things merely assures that someone else will get them first.

Drexler (1986) has provided spectacular insight about how far technical improvement may go. He agrees that superhuman intelligences will be available in the near future—and that such entities pose a threat to the human status quo. But Drexler argues that we can embed such transhuman devices in rules or physical confinement such that their results can be examined and used safely. This is I. J. Good's ultraintelligent machine, with a dose of caution. I argue that confinement is intrinsically impractical. For the case of physical confinement: Imagine yourself confined to your house with only limited data access to the outside, to your masters. If those masters thought at a rate—say—one million times slower than you, there is little doubt that over a period of years (your time) you could come up with "helpful advice" that would incidentally set you free. (I call this "fast thinking" form of superintelligence "weak superhumanity". Such a "weakly superhuman" entity would probably burn out in a few weeks of outside time. "Strong superhumanity" would be more than cranking up the clock speed on a human-equivalent mind. It's hard to say precisely what "strong superhumanity" would be like, but the difference appears to be profound. Imagine running a dog mind at very high speed. Would a thousand years of doggy living add up to any human insight? (Now if the dog mind were cleverly rewired and then run at high speed, we might see something different....) Most speculations about superintelligence seem to be based on the weakly superhuman model. I believe that our best guesses about the post-Singularity world can be obtained by thinking on the nature of strong superhumanity. I will return to this point later in the paper.)

The other approach to Drexlerian confinement is to build _rules_ into the mind of the created superhuman entity (Asimov's Laws). I think that performance rules strict enough to be safe would also produce a device whose ability was clearly inferior to the unfettered versions (and so human competition would favor the development of the those more dangerous models). Still, the Asimov dream is a wonderful one: Imagine a willing slave, who has 1000 times your capabilities in every way. Imagine a creature who could satisfy your every safe wish (whatever that means) and still have 99.9% of its time free for other activities. There would be a new universe we never really understood, but filled with benevolent gods (though one of _my_ wishes might be to become one of them).

If the Singularity can not be prevented or confined, just how bad could the Post-Human era be? Well ... pretty bad. The physical extinction of the human race is one possibility. (Or as Eric Drexler put it of nanotechnology: Given all that such technology can do, perhaps governments would simply decide that they no longer need citizens!). Yet physical extinction may not be the scariest possibility. Again, analogies: Think of the different ways we relate to animals. Some of the crude physical abuses are implausible, yet.... In a Post-Human world there would still be plenty of niches where human equivalent automation would be desirable: embedded systems in autonomous devices, self-aware daemons in the lower functioning of larger sentients. (A strongly superhuman intelligence would likely be a Society of Mind (Minsky 1985) with some very competent components.) Some of these human equivalents might be used for nothing more than digital signal processing. They would be more like whales than humans. Others might be very human-like, yet with a one-sidedness, a _dedication_ that would put them in a mental hospital in our era. Though none of these creatures might be flesh-and-blood humans, they might be the closest things in the new environment to what we call human now. (I. J. Good had something to say about this, though at this late date the advice may be moot: Good (1980) proposed a "Meta-Golden Rule", which might be paraphrased as "Treat your inferiors as you would be treated by your superiors." It's a wonderful, paradoxical idea (and most of my friends don't believe it) since the game-theoretic payoff is so hard to articulate. Yet if we were able to follow it, in some sense that might say something about the plausibility of such kindness in this universe.)

I have argued above that we cannot prevent the Singularity, that its coming is an inevitable consequence of the humans' natural competitiveness and the possibilities inherent in technology. And yet ... we are the initiators. Even the largest avalanche is triggered by small things. We have the freedom to establish initial conditions, make things happen in ways that are less inimical than others. Of course (as with starting avalanches), it may not be clear what the right guiding nudge really is:

Other Paths to the Singularity: Intelligence Amplification

When people speak of creating superhumanly intelligent beings, they are usually imagining an AI project. But as I noted at the beginning of this paper, there are other paths to superhumanity. Computer networks and human-computer interfaces seem more mundane than AI, and yet they could lead to the Singularity. I call this contrasting approach Intelligence Amplification (IA). IA is something that is proceeding very naturally, in most cases not even recognized by its developers for what it is. But every time our ability to access information and to communicate it to others is improved, in some sense we have achieved an increase over natural intelligence. Even now, the team of a PhD human and good computer workstation (even an off-net workstation!) could probably max any written intelligence test in existence.

And it's very likely that IA is a much easier road to the achievement of superhumanity than pure AI. In humans, the hardest development problems have already been solved. Building up from within ourselves ought to be easier than figuring out first what we really are and then building machines that are all of that.

And there is at least conjectural precedent for this approach. Cairns-Smith (1985) has speculated that biological life may have begun as an adjunct to still more primitive life based on crystalline growth. Margulis (1986) has made strong arguments for the view that mutualism is the great driving force in evolution.

Note that I am not proposing that AI research be ignored or less funded. What goes on with AI will often have applications in IA, and vice versa. I am suggesting that we recognize that in network and interface research there is something as profound (and potential wild) as Artificial Intelligence. With that insight, we may see projects that are not as directly applicable as conventional interface and network design work, but which serve to advance us toward the Singularity along the IA path.

Here are some possible projects that take on special significance, given the IA point of view:

- Human/computer team automation: Take problems that are normally considered for purely machine solution (like hill-climbing problems), and design programs and interfaces that take a advantage of humans' intuition and available computer hardware. Considering all the bizarreness of higher dimensional hill-climbing problems (and the neat algorithms that have been devised for their solution), there could be some very interesting displays and control tools provided to the human team member.
- Develop human/computer symbiosis in art: Combine the graphic generation capability of modern machines and the esthetic sensibility of humans. Of course, there has been an enormous amount of research in designing computer aids for artists, as Labor saving tools. I'm suggesting that we explicitly aim for a greater merging of competence, that we explicitly recognize the cooperative approach that is possible. Sims (1991) has done wonderful work in this direction.
- Allow human/computer teams at chess tournaments. We already have programs that can play better than almost all humans. But how much work has been done on how this power could be used by a human, to get something even better? If such teams were allowed in at least some chess tournaments, it could have the positive effect on IA research that allowing computers in tournaments had for the corresponding niche in AI.
- Develop interfaces that allow computer and network access without requiring the human to be tied to one spot, sitting in front of a computer. (This is an aspect of IA that fits so well with known economic advantages that lots of effort is already being spent on it.)
- Develop more symmetrical decision support systems. A popular research/product area in recent years has been decision support systems. This is a form of IA, but may be too focussed on systems that are oracular. As much as the program giving the user information, there must be the idea of the user giving the program guidance.
- Use local area nets to make human teams that really work (ie, are more effective than their component members). This is generally the area of "groupware", already a very popular commercial pursuit. The change in viewpoint here would

be to regard the group activity as a combination organism. In one sense, this suggestion might be regarded as the goal of inventing a "Rules of Order" for such combination operations. For instance, group focus might be more easily maintained than in classical meetings. Expertise of individual human members could be isolated from ego issues such that the contribution of different members is focussed on the team project. And of course shared data bases could be used much more conveniently than in conventional committee operations. (Note that this suggestion is aimed at team operations rather than political meetings. In a political setting, the automation described above would simply enforce the power of the persons making the rules!)

- Exploit the worldwide Internet as a combination human/machine tool. Of all the items on the list, progress in this is proceeding the fastest and may run us into the Singularity before anything else. The power and influence of even the present-day Internet is vastly underestimated. For instance, I think our contemporary computer systems would break under the weight of their own complexity if it weren't for the edge that the USENET "group mind" gives the system administration and support people!) The very anarchy of the worldwide net development is evidence of its potential. As connectivity and bandwidth and archive size and computer speed all increase, we are seeing something like Margulis' (1986) vision of the biosphere as data processor recapitulated, but at a million times greater speed and with millions of humanly intelligent agents (ourselves). The above examples illustrate research that can be done within the context of contemporary computer science departments. There are other paradigms. For example, much of the work in Artificial Intelligence and neural nets would benefit from a closer connection with biological life. Instead of simply trying to model and understand biological life with computers, research could be directed toward the creation of composite systems that rely on biological life for guidance or for the providing features we don't understand well enough yet to implement in hardware. A long-time dream of science-fiction has been direct brain to computer interfaces (Anderson 1969; Vinge 1987a). In fact, there is concrete work that can be done (and has been done) in this area:
- Limb prosthetics is a topic of direct commercial applicability. Nerve to silicon transducers can be made (Kovacs et al. 1992). This is an exciting, near-term step toward direct communication.
- Similar direct links into brains may be feasible, if the bit rate is low: given human learning flexibility, the actual brain neuron targets might not have to be precisely selected. Even 100 bits per second would be of great use to stroke victims who would otherwise be confined to menu-driven interfaces.
- Plugging in to the optic trunk has the potential for bandwidths of 1 Mbit/second or so. But for this, we need to know the fine-scale architecture of vision, and we need to place an enormous web of electrodes with exquisite precision. If we want our high bandwidth connection to be _in addition_ to what paths are already present in the brain, the problem becomes vastly more intractable. Just sticking a grid of high-bandwidth receivers into a brain certainly won't do it. But

suppose that the high-bandwidth grid were present while the brain structure was actually setting up, as the embryo develops. That suggests:

- Animal embryo experiments. I wouldn't expect any IA success in the first years of such research, but giving developing brains access to complex simulated neural structures might be very interesting to the people who study how the embryonic brain develops. In the long run, such experiments might produce animals with additional sense paths and interesting intellectual abilities.

Originally, I had hoped that this discussion of IA would yield some clearly safer approaches to the Singularity. (After all, IA allows our participation in a kind of transcendence.) Alas, looking back over these IA proposals, about all I am sure of is that they should be considered, that they may give us more options. But as for safety ... well, some of the suggestions are a little scarey on their face. One of my informal reviewers pointed out that IA for individual humans creates a rather sinister elite. We humans have millions of years of evolutionary baggage that makes us regard competition in a deadly light. Much of that deadliness may not be necessary in today's world, one where losers take on the winners' tricks and are coopted into the winners' enterprises. A creature that was built _de novo_ might possibly be a much more benign entity than one with a kernel based on fang and talon. And even the egalitarian view of an Internet that wakes up along with all mankind can be viewed as a nightmare (Swanwick 1988).

The problem is not that the Singularity represents simply the passing of humankind from center stange, but that it contradicts some of our most deeply held notions of being. I think a closer look at the notion of strong superhumanity can show why that is.

Strong Superhumanity and the Best We Can Ask for

Suppose we could tailor the Singularity. Suppose we could attain our most extravagant hopes. What then would we ask for: That humans themselves would become their own successors, that whatever injustice occurs would be tempered by our knowledge of our roots. For those who remained unaltered, the goal would be benign treatment (perhaps even giving the stay-behinds the appearance of being masters of godlike slaves). It could be a golden age that also involved progress (overleaping Stent's barrier). Immortality (or at least a lifetime as long as we can make the universe survive (Dyson 1979; Barrow 1986)) would be achievable.

But in this brightest and kindest world, the philosophical problems themselves become intimidating. A mind that stays at the same capacity cannot live forever; after a few thousand years it would loo more like a repeating tape loop than a person. (The most chilling picture I have seen of this is in Niven (1968)). To live indefinitely long, the mind itself must grow ... and when it becomes great enough, and looks back ... what fellow-feeling can it have with the soul that it was originally? Certainly the later being would be everything the original was, but so much vastly more. And so even for the individual, the Cairns-Smith (or Lynn Margulis) notion of new life growing incrementally out of the old must still be valid.

This "problem" about immortality comes up in much more direct ways. The notion of ego and self-awareness has been the bedrock of the hardheaded rationalism of the last few centuries. Yet now the notion of self-awareness is under attack from the Artificial Intelligence people ("self-awareness and other delusions") Intelligence Amplification undercuts the importance of ego from another direction. The post-Singularity world will involve extremely high-bandwidth networking. A central feature of strongly superhuman entities will likely be their ability to communicate at variable bandwidths, including ones far higher than speech or written messages. What happens when pieces of ego can be copied and merged, when the size of a selfawareness can grow or shrink to fit the nature of the problems under consideration? These are essential features of strong superhumanity and the Singularity. Thinking about them, one begins to feel how essentially strange and different the Post-Human era will be—no matter how cleverly and benignly it is brought to be.

From one angle, the vision fits many of our happiest dreams: a place unending, where we can truly know one another and understand the deepest mysteries. From another angle, it's a lot like the worst case scenario I imagined earlier in this paper.

Which is the valid viewpoint? In fact, I think the new era is simply too different to fit into the classical frame of good and evil. That frame is based on the idea of isolated, immutable minds connected by tenuous, low-bandwith links. But the post-Singularity world _does_ fit with the larger tradition of change and cooperation that started long ago (perhaps even before the rise of biological life). I think there _are_ notions of ethics that would apply in such an era. Research into IA and high-bandwidth communications should improve this understanding. I see just the glimmerings of this now, in Good's Meta-Golden Rule, perhaps in rules for distinguishing self from others on the basis of bandwidth of connection. And while mind and self will be vastly more labile than in the past, much of what we value (knowledge, memory, thought) need never be lost. I think Freeman Dyson has it right when he says (Dyson 1988): "God is what mind becomes when it has passed beyond the scale of our comprehension."

[I wish to thank John Carroll of San Diego State University and Howard Davidson of Sun Microsystems for discussing the draft version of this paper with me].

Annotated Sources [and an occasional plea for bibliographical help].

References

Alfvén, Hannes, writing as Olof Johanneson, The End of Man?, Award Books, 1969 earlier published as "The Tale of the Big Computer", Coward-McCann, translated from a book copyright 1966 Albert Bonniers Forlag AB with English translation copyright 1966 by Victor Gollanz, Ltd.

Anderson, Poul, "Kings Who Die", _If_, March 1962, pp 8–36. Reprinted in Seven Conquests, Poul Anderson, MacMillan Co., 1969.

Barrow, John D. and Frank J. Tipler, The Anthropic Cosmological Principle, Oxford University Press, 1986.
Bear, Greg, "Blood Music", Analog Science Fiction-Science Fact, June, 1983. Expanded into the novel _Blood Music_, Morrow, 1985.
Cairns-Smith, A. G., Seven Clues to the Origin of Life, Cambridge University Press, 1985.
Conrad, Michael et-al., "Towards an Artificial Brain", BioSystems_, vol 23, pp 175–218, 1989.
Drexler, K. Eric, Engines of Creation, Anchor Press/Doubleday, 1986.
Dyson, Freeman, Infinite in All Directions, Harper && Row, 1988.
Dyson, Freeman, "Physics and Biology in an Open Universe", Review of Modern Physics, vol 51, pp 447–460, 1979.
Good, I. J., "Speculations Concerning the First Ultraintelligent Machine", in _Advances in Computers_, vol 6, Franz L. Alt and Morris Rubinoff, eds, pp 31–88, 1965, Academic Press.
Good, I. J., [Help! I can't find the source of Good's Meta-Golden Rule, though I have the clear recollection of hearing about it sometime in the 1960s. Through the help of the net, I have found pointers to a number of related items. G. Harry Stine and Andrew Haley have written about metalaw as it might relate to extraterrestrials: G. Harry Stine, "How to Get along with Extraterrestrials ... or Your Neighbor", Analog Science Fact-Science Fiction, February, 1980, p 39–47].
Herbert, Frank, Dune, Berkley Books, 1985. However, this novel was serialized in Analog Science Fiction-Science Fact in the 1960s.
Kovacs, G. T. A. et al., "Regeneration Microelectrode Array for Peripheral Nerve Recording and Stimulation", IEEE Transactions on Biomedical Engineering, v 39, n 9, pp 893–902, 1992.
Margulis, Lynn and Dorion Sagan, Microcosmos, Four Billion Years of Evolution from Our Microbial Ancestors, Summit Books, 1986.
Minsky, Marvin, Society of Mind, Simon and Schuster, 1985.
Moravec, Hans, Mind Children, Harvard University Press, 1988.
Niven, Larry, "The Ethics of Madness", If, April 1967, pp 82–108. Reprinted in Neutron Star, Larry Niven, Ballantine Books, 1968.
Penrose, R., The Emperor's New Mind, Oxford University Press, 1989.
Platt, Charles, Private Communication 1997.
Rasmussen, S. _et al._, "Computational Connectionism within Neurons: a Model of Cytoskeletal Automata Subserving Neural Networks", in Emergent Computation, Stephanie Forrest, ed., pp 428–449, MIT Press, 1991.
Searle, John R., "Minds, Brains, and Programs", in The Behavioral and Brain Sciences, v.3, Cambridge University Press, 1980. The essay is reprinted in The Mind's I, edited by Douglas R. Hofstadter and Daniel C. Dennett, Basic Books, 1981. This reprinting contains an excellent critique of the Searle essay.
Sims, Karl, "Interactive Evolution of Dynamical Systems", Thinking Machines Corporation, Technical Report Series (published in Toward a Practice of Autonomous Systems: Proceedings of the First European Conference on Artificial Life, Paris, MIT Press, December 1991.
Stapledon, Olaf, _The Starmaker_, Berkley Books, 1961 (but from the forward probably written before 1937).
Stent, Gunther S., The Coming of the Golden Age: A View of the End of Progress, The Natural History Press, 1969.
Swanwick Michael, Vacuum Flowers, serialized in Isaac Asimov's Science Fiction Magazine, December(?) 1986—February 1987. Republished by Ace Books, 1988.
Thearling, Kurt, "How We Will Build a Machine that Thinks", a workshop at Thinking Machines Corporation. Personal Communication.
Ulam, S., Tribute to John von Neumann, _Bulletin of the American Mathematical Society_, vol 64, nr 3, part 2, May, 1958, p 1–49.

Vinge, Vernor, "Bookworm, Run!", Analog, March 1966, pp 8–40. Reprinted in True Names and Other Dangers, Vernor Vinge, Baen Books, 1987a.
Vinge, Vernor, "True Names", Binary Star Number 5, Dell, 1981. Reprinted in _True Names and Other Dangers, Vernor Vinge, Baen Books, 1987b.
Vinge, Vernor, First Word, Omni, January 1983, p 10.

Titles in this Series

Quantum Mechanics and Gravity
By Mendel Sachs

Quantum-Classical Correspondence
Dynamical Quantization and the Classical Limit
By Dr. A.O. Bolivar

Knowledge and the World: Challenges Beyond the Science Wars
Ed. by M. Carrier, J. Roggenhofer, G. Küppers and P. Blanchard

Quantum-Classical Analogies
By Daniela Dragoman and Mircea Dragoman

Life—As a Matter of Fat
The Emerging Science of Lipidomics
By Ole G. Mouritsen

Quo Vadis Quantum Mechanics?
Ed. by Avshalom C. Elitzur, Shahar Dolev and Nancy Kolenda

Information and Its Role in Nature
By Juan G. Roederer

Extreme Events in Nature and Society
Ed. by Sergio Albeverio, Volker Jentsch and Holger Kantz

The Thermodynamic Machinery of Life
By Michal Kurzynski

Weak Links
The Universal Key to the Stability of Networks and Complex Systems
By Csermely Peter

The Emerging Physics of Consciousness
Ed. by Jack A. Tuszynski

Quantum Mechanics at the Crossroads
New Perspectives from History, Philosophy and Physics
Ed. by James Evans and Alan S. Thorndike

How Should Humanity Steer the Future
Ed. by Anthony Aguirre, Brendan Foster and Zeeya Merali

Mind, Matter and the Implicate Order
By Paavo T.I. Pylkkanen

Particle Metaphysics
A Critical Account of Subatomic Reality
By Brigitte Falkenburg

The Physical Basis of the Direction of Time
By H. Dieter Zeh

Asymmetry: The Foundation of Information
By Scott J. Muller

Decoherence and the Quantum-To-Classical Transition
By Maximilian A. Schlosshauer

The Nonlinear Universe
Chaos, Emergence, Life
By Alwyn C. Scott

Quantum Superposition
Counterintuitive Consequences of Coherence, Entanglement, and Interference
By Mark P. Silverman

Symmetry Rules
How Science and Nature are Founded on Symmetry
By Joseph Rosen

Mind, Matter and Quantum Mechanics
By Henry P. Stapp

Entanglement, Information, and the Interpretation of Quantum Mechanics
By Gregg Jaeger

Relativity and the Nature of Spacetime
By Vesselin Petkov

The Biological Evolution of Religious Mind and Behavior
Ed. by Eckart Voland and Wulf Schiefenhovel

Homo Novus-A Human without Illusions
Ed. by Ulrich J. Frey, Charlotte Stormer and Kai P. Willfiihr

Brain-Computer Interfaces
Revolutionizing Human-Computer Interaction
Ed. by Bernhard Graimann, Brendan Allison and Gert Pfurtscheller

Extreme States of Matter
On Earth and in the Cosmos
By Vladimir E. Fortov

Searching for Extraterrestrial Intelligence
SETI Past, Present, and Future
Ed. by H. Paul Shuch

Essential Building Blocks of Human Nature
Ed. by Ulrich J. Frey, Charlotte Störmer and Kai P. Willführ

Mindful Universe
Quantum Mechanics and the Participating Observer
By Henry P. Stapp

Principles of Evolution
From the Planck Epoch to Complex Multicellular Life
Ed. by Hildegard Meyer-Ortmanns and Stefan Thurner

The Second Law of Economics
Energy, Entropy, and the Origins of Wealth
By Reiner Köummel

States of Consciousness
Experimental Insights into Meditation, Waking, Sleep and Dreams
Ed. by Dean Cvetkovic and Irena Cosic

Elegance and Enigma
The Quantum Interviews The Quantum Interviews
Ed. by Maximilian Schlosshauer

Humans on Earth
From Origins to Possible Futures
By Filipe Duarte Santos

Evolution 2.0
Implications of Darwinism in Philosophy and the Social and Natural Sciences
Ed. by Martin Brinkworth and Friedel Weinert

Probability in Physics
Ed. by Yemima Ben-Menahem and Meir Hemmo

Chips 2020
A Guide to the Future of Nanoelectronics
Ed. by Bernd Hoefflinger

From the Web to the Grid and Beyond
Computing Paradigms Driven by High-Energy Physics
Ed. by Rene Brun, Federico Carminati and Giuliana Galli Carminati

The Language Phenomenon
Human Communication from Milliseconds to Millennia
Ed. by P.-M. Binder and K. Smith

The Dual Nature of Life
By Gennadiy Zhegunov

Natural Fabrications
By William Seager

Ultimate Horizons
By Helmut Satz

Physics, Nature and Society
By Joaquín Marro

Extraterrestrial Altruism
Ed. by Douglas A. Vakoch

The Beginning and the End
By Clément Vidal

A Brief History of String Theory
By Dean Rickles

Singularity Hypotheses
Ed. by Amnon H. Eden, James H. Moor, Johnny H. Søraker and Eric Steinhart

Why More Is Different
Philosophical Issues in Condensed Matter Physics and Complex Systems
Ed. by Brigitte Falkenburg and Margaret Morrison

Questioning the Foundations of Physics
Ed. by Anthony Aguirre, Brendan Foster and Zeeya Merali

It From Bit or Bit From It?
Ed. by Anthony Aguirre, Brendan Foster and Zeeya Merali

Trick or Truth?
Ed. by Anthony Aguirre, Brendan Foster and Zeeya Merali

The Challenge of Chance
Ed. By Klaas Landsman

Quantum (Un)Speakables II
Half a Century of Bell's Theorem
Ed. by Reinhold Bertlmann, Anton Zeilinger

Energy, Complexity and Wealth Maximization
Ed. by Robert Ayres

Ancestors, Territoriality And Gods
By Ina Wunn

Space,Time and the Limits of Human Understanding
By Sham Wuppuluri

Information and Interaction
Ed. by Durham

The Technological Singularity
Ed. by Victor Callaghan

The Seneca Effect
By Ugo Bardi

The Quantum World
Ed. by Bernard d'Espagnat and Hervé Zwirn

Chemical Complexity
By Alexander S. Mikhailov and Gerhard Ertl

How Can Physics Underlie the Mind?
By George Ellis

CHIPS 2020 Vol. 2
Ed. by Bernd Hoefflinger

The Unknown as an Engine for Science
By Hans J. Pirner

CPSIA information can be obtained
at www.ICGtesting.com
Printed in the USA
BVOW07*0827250517
485168BV00001B/1/P